Cities and Nature

Cities and Nature illustrates how the city is part of the environment, and how it is subject to environmental constraints and opportunities. The city has been treated in geographical writings as only a social phenomena, and at the same time, environmental scientists have tended to ignore the urban. This book aims to reconnect the science and social science through the examination of the urban. It critiques the dominant academic discourse which ignores the environmental base of urban life and living, and discusses the urban natural environment and how this is subjected to social influences.

The book is organized around three central themes: 1) urban environment in historical context; 2) issues in urban–nature relations and; 3) realigning urban–nature relations. It discusses ideas such as pollution as a physical environmental fact, often created or impacted by economic, cultural and political changes. Pollution, for example, is viewed, as a social act: consuming patterns of everyday activities–driving, showering, shopping, eating–have environmental impacts. The authors seek to reintroduce a social science perspective in examining urban nature, the city and its physical environment.

Cities and Nature clearly illustrates the physical and social elements of the urban environment and shows how these are important to examining the city. It includes further reading and boxed case studies on Bangladesh, Paris, Delhi, Rome, Cubatao, Thailand, Los Angeles, Chicago, New Orleans and Toronto. This book will be an asset to students and researchers in Environmental Studies, Urban Studies and Planning.

Lisa Benton-Short is an Associate Professor of Geography at The George Washington University in Washington, DC. An urban geographer, she has research interests in environmental issues in cities, parks and public spaces, and monuments and memorials. She is also interested in globalization, urbanization and migration.

John Rennie Short is Professor of Geography and Public Policy, University of Maryland, Baltimore County. He has published twenty-eight books and numerous articles and is recognized as an international authority on the study of cities.

Routledge critical introductions to urbanism and the city

Edited by Malcolm Miles, University of Plymouth, UK
and John Rennie Short, University of Maryland, USA

International Advisory Board:

Franco Bianchini	Jane Rendell
Kim Dovey	Saskia Sassen
Stephen Graham	David Sibley
Tim Hall	Erik Swyngedouw
Phil Hubbard	Elizabeth Wilson
Peter Marcuse	

The series is designed to allow undergraduate readers to make sense of, and find a critical way into, urbanism. It will:

- Cover a broad range of themes
- Introduce key ideas and sources
- Allow the author to articulate her/his own position
- Introduce complex arguments clearly and accessibly
- Bridge disciplines, and theory and practice
- Be affordable and well designed

The series covers social, political, economic, cultural and spatial concerns. It will appeal to students in architecture, cultural studies, geography, popular culture, sociology, urban studies, urban planning. It will be trans-disciplinary. Firmly situated in the present, it also introduces material from the cities of modernity and post-modernity.

Published:
Cities and Consumption – Mark Jayne
Cities and Cultures – Malcolm Miles
Cities and Nature – Lisa Benton-Short and John Rennie Short

Forthcoming:
Cities and Economies – Yeong- Hyun Kim and John Rennie Short
Cities and Cinema – Barbara Mennel
Cities, Politics and Power – Simon Parker
Urban Erotics – David Bell and John Binnie
Children Youth and the City – Kathrin Horshelmann and Lorraine van Blerk
Cities and Gender – Helen Jarvis, Jonathan Cloke & Paula Kantor

Cities and Nature

*By Lisa Benton-Short and
John Rennie Short*

Routledge
Taylor & Francis Group

LONDON AND NEW YORK

First published 2008
By Routledge
2 Park Square, Milton Park, Abingdon, Oxon, OX14 4RN

Simultaneously published in the USA and Canada
By Routledge
270 Madison Avenue, New York, NY 10016

Reprinted 2009 (twice)

Routledge is an imprint of the Taylor & Francis Group, an informa business

© 2008 Lisa Benton-Short and John Rennie Short

Typeset in Times New Roman by Keyword Group Ltd.
Printed and bound in Great Britain by TJ International Ltd, Padstow, Cornwall

British Library Cataloguing in Publication Data
A catalogue record for this book is available from the British Library

Library of Congress Cataloging in Publication Data
Benton-Short, Lisa.
Cities and nature / by Lisa Benton-Short and John Rennie Short.
p. cm.
Includes bibliographical references and index.
ISBN 978-415-35588-9 (hard cover) – ISBN 978-0-415-35589-6
(soft cover) 1. Urban ecology. 2. Urbanization—Environmental aspects.
3. City planning—Environmental aspects. 4. Urban pollution.
5. Sustainable development. I. Short, John R. II. Title.
HT241.B46 2007
307.76—dc22   2007020942

ISBN 10: 0-415-35589-3(pbk)
ISBN 10: 0-415-35588-5(hbk)
ISBN 10: 0-203-00232-6(ebk)

ISBN 13: 978-0-415-35589-6(pbk)
ISBN 13: 978-0-415-35588-9(hbk)
ISBN 13: 978-0-203-00232-2(ebk)

This book is dedicated to Bonnie and Harriet,
who have given us unconditional love
and support.

Contents

Figures

Tables

Acknowledgements

We would like to first acknowledge that this book is the result of numerous years of teaching the topic of "urban environmental issues" to undergraduates and graduates at Colgate University, The George Washington University and University of Maryland Baltimore County. In the process of teaching this course, we found our students were invaluable in asking provocative questions and providing new and inspirational sources of material in their term papers and research projects.

Our thanks also go to the three anonymous reviewers who provided thoughtful and important suggestions to improve this book.

We would also like to thank Nuala Cowen, Director of the Spatial Analysis Lab at The George Washington University for her assistance in making several of the maps, graphs and charts included in this book. Thanks go to Nikolas Schiller, who provided the cover image for the book. The image is an aerial photograph of Central Park in New York City, which Nikolas "tessellated" and to which he applied a hexagon quilt projection. For more information about his cartographic artwork go to http://www.nikolasrschiller.com or http://www.geospatialart.com We would also like to thank the following for granting permission to reproduce images in this work: Stefanie and Katie Garry, Jesse Goldman, undergraduate students at The George Washington University; colleagues Elizabeth Chacko, David Rain and Joe Dymond; Rob Crandall a professional photographer and friend; and our best friend Michele Antoinette Judd.

Every effort has been made to contact copyright holders for their permission to reprint material in this book. The publishers would be grateful to hear from any copyright holder who is not here acknowledged, and will undertake to rectify any errors or omissions in future editions of this book.

Lastly, we appreciate Cosmo, Rockie and Columbus, who sat beside us in chairs and on desk tops, and Murphy, who reminded us when it was time to stop and take a walk.

Part I
The Urban Environment in History

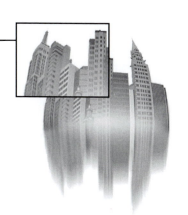

1 The city and nature

In the summer of 1995 a heat wave struck the city of Chicago. For over a week in July the temperature reached over 100 degrees every day. By July 20 over 700 people had died. It was referred to in the press as a "natural disaster", the unfortunate outcome of a freak meteorological condition. In his 2002 book *Heat Wave*, Eric Klinenberg undertook a social autopsy of the event. He found that deaths were greatest amongst more elderly people living on their own. The tragedy was not simply a natural disaster but the outcome of the social isolation of seniors, retrenchment of public assistance and declining neighborhoods. Most victims were seniors who lived alone in neighborhoods that lacked a sense of community and where there was perception of danger in the streets. Trapped inside their homes, and with few visits from public health officials, many poor isolated seniors overheated and died. The disaster was not the result of high temperatures but high temperatures as mediated through a complex set of social and political relationships.[1]

We take two points from the story. First, the city is part of the ecosystem, in this case part of the rising temperatures of summer warming. Second, these environmental processes are filtered through social arrangements of political and economic differences. There are no such things as "natural" disasters in cities just as there are no cities independent of nature.

We need to make the point that "natural" disasters are, on closer inspection, much more closely connected to social processes than we often acknowledge. When people build on eroding hillsides or locate houses in earthquake zones then the natural disaster turns out to be in part a social construction. And the term "natural" disaster also hides the social-economic implication of their effect. Citizens of poor countries are more affected by floods and storms because they do not have expensive technology to provide as much early warning or as rapid evacuation as richer countries. The same storm will have vastly different human consequences in different places. And even within the same city the experience of disaster by rich and poor residents can be vastly different. The searing images of New Orleans in the immediate aftermath of Hurricane Katrina remind us of the gulf that separates those able to leave and those stranded in the city.

Disasters provide a visible connection between nature and the city. Upon closer inspection the natural appears more social and the social life of cities is more accurately seen as implicated in environmental processes. The city is the center of a society–environment dialectic. In this introductory chapter we will explore three general notions: the city as ecosystem, nature in the city and urban sustainability. These themes run throughout all of the subsequent chapters.

Words and definitions

Words are important, slippery, relational things. We need to make clear what we mean when we use the terms nature, environment and city. Nature has many different meanings but two of the most important are: essential quality (as in "it is in his nature to do these things"); the material world that can include or exclude human beings. Words are relational and nature was often counterpoised against technology and/or culture. In this book we will use the term "nature" to refer to the material world that includes human beings. We use it less in opposition to human society and more as the container of physical resources and cultural meanings. We will use nature and environment interchangeably.

There has long been a distinction between nature and city in linguistic usage. The main point of this book is to show that the city is part of the material world and this materiality is shaped and structured by and in cities.

The intellectual background

Cities provide an inevitable contrast to the "natural". A consistent strand of thought has sought to place the city as a human invention in opposition to the "natural", the "pristine", and the "wilderness". Protecting the environment has usually meant halting the encroachment into pristine areas such as rainforests and tundra. Most often, environmental protection has been defined as meaning something outside of, and mostly unrelated to, the concerns and interests of our cities. Cities have been described and understood as somehow separate from the so-called "natural world". This has been reinforced by the appearance of an increased separation of life in the city from the wider environmental context.

When food is more available in a supermarket aisle instead of in the fields outside our homes and when we can turn up the heating to keep out the cold or turn on the air conditioning to keep out the oppressive heat there is a tendency to see the city as somehow removed and independent from the physical world.

Urban theorizing has for a long time been conducted as if a city was on a flat, featureless plain. Urban studies have long ignored the physical nature of cities; instead, the emphasis has been on the social, political and economic rather than the ecological. And yet cities are ecological systems, they are predicated upon the physical world as mediated through the complex prism of social and economic power. In recent years there has been a renewal of interest in the city as an ecological system with emphasis on the complex relationships between environmental issues and urban concerns, and between social networks and ecosystem flows.[2] To take just one example: William Solecki and Cynthia Rozenzweig look at the biodiversity-urban society relationships in the greater New York Metropolitan Region.[3] They use such concepts as the "ecological footprint" and vulnerability to global environmental change to analyze the current interactions between biodiversity and urban society.

In this new exciting body of urban ecological work, cities are now seen as much natural as wilderness and the environment as much social as the city. The city is implicated in the "natural" world in connections that embody and reflect social, economic and political power. The city is an integral part of nature and nature is intimately interwoven into the social life of cities.

The city as ecosystem

Physical geographer Ian Douglas suggested that the city itself can be seen as an ecosystem with inputs of energy and water and outputs of noise, climate change, sewage, garbage and air pollutants.[4] Another way to consider the city–nature dialectic then is to consider the city as an ecological system with a measurable amount of environmental inputs and outputs. Amongst the most obvious inputs are energy and water (see Table 1.1).

Human activity in the city is dependent on large and consistent inputs of energy. When we leave heated buildings to drive in cars to purchase goods we use energy. The commercial activities we pursue and the microclimates we create (heating in winter, cooling in summer) all use energy. In seeking to overthrow the tyranny of nature, cities use prodigious amounts of energy. Cities are deeply dependent on energy sources. In the US, since the beginning of the twentieth century, petroleum has traditionally been very cheap and cities now sprawl across the landscape. In countries where energy is more expensive, cities tend to be higher

Table 1.1 *Resources in and out of London, 2000*

Resources in	Tons/year
Oxygen	40,000,000
Water	876,000,000,000 litres
Food	6,900,000
Paper	2,200,000
Plastic	2,100,000
Construction material	27,000,000
(bricks, sand, concrete)	
Energy needs (tons of oil)	13,276,000

Resources out	Tons/year
Carbon Dioxide	41,000,000
Sulfur Dioxide	400,000
Nitrogen Oxides	280,000
Sewage and sludge	7,500,000
Industrial and Commercial Waste	14,029,000
Household Waste	3,900,000

Source: The City Limits project: a resource flow and ecological footprint analysis of Greater London at www.citylimitslondon.com

in density and more reliant on public transport. Large-scale suburban sprawl is a function of cheap energy. It is tempting to theorize the impact of a long-term, sustained increase in energy prices on suburban sprawl and urban structure.

Water is an essential ingredient of life. The people and commerce of cities are utterly dependent upon water. One of the largest urban differences in the world is between cities with clean, easily accessible water and others with expensive, inaccessible and polluted water supplies. In order to provide inexpensive and clean water immense engineering projects have been undertaken. And as cities have grown the catchments areas have extended outwards and the engineering sophistication of piping in water has grown and deepened. In poorer cities polluted urban water remains a major source of disease and illness especially for children.

Even in rich countries the availability of fresh water is a determinant of the limits of urban growth. In the arid west of the US, for example, urban growth has been predicated upon massive federal subsidies and expensive engineering projects that have provided fresh water at low cost to the consumers. The ecological limits are always more flexible than the environmental determinists suggest but they are not infinitely extendable. We may be reaching the "water" limits of urban growth in the arid US.

Cities also modify the environment. The most obvious example of this is the urban heat island. Cities tend to be warmer because of the amount of extra heat produced in the city and the heat absorption of man-made materials such as tarmac, asphalt and concrete. Heat is absorbed by these surfaces during the day and released at night. The net result is for the air around cities to be warmer than surrounding rural areas. One side effect is to reduce the need for heating in the winter but to increase the need for air-conditioning in the summer. The heat island means you can turn the heating down in London in December but need to increase the air conditioning in Washington, DC in August. The extra heat causes a thermally induced upward movement of air, and an increase in cloud and raindrop formation. Cities are often cloudier, more prone to thunder and slightly warmer than surrounding rural areas.

Human activity in the city also produces pollutants. Industrial processes and auto engines emit substances that include carbon oxides, sulfur oxides, hydrocarbons, dust, soot and lead. The air in cities has traditionally been very unhealthy, which is part of the reason for the higher urban death rate throughout most of human history. The pall of smog that hangs over many cities is a visible reminder of the effects of concentrated human activity on the environment. The pollutants of cities are not only injurious to the health of individuals, but they also cause more general damage; cities are in part a major cause of global warming and ozone depletion.

A major output of cities is garbage. High mass consumption in association with elaborate packaging has created a rising tide of garbage in cities. Burning it causes air pollution, while hiding it leads to massive landfills. The environmental justice literature shows that many environmentally hazardous facilities are generally located in poor, minority and more weakly organized communities. Issues of environmental management are tied in to wider issues of equity and social justice. Patterns of environmental racism are clear when we note that most noxious facilities are located in lower income, more marginal communities.

Cities also emit noise. Cities are noisy places and households who inhabit busy urban streets for more than 15 years are on average likely to experience a 50 percent reduction in hearing capacity. The effects of noise pollution vary from annoyance to deterioration in hearing. A high backroom noise level leads to a general increase in stress and the lessening of the quality of urban life.

Cities are an integral part of the hydrological cycle. Cities impact the daily and seasonal flows of water. The large amount of impermeable surfaces, for example, means that when it rains run off levels spike dramatically. Cities thus need to create modified flows through channels and conduits that can cope with the irregular high flow rates. But the large amount of impermeable surfaces in association with the channelization of water courses can lead to distinct surges in water flow after rain and in many cases patterns of flooding. As urbanization increases so to does the overloading of the hydrological cycle. Cities also modify the flow and direction of rivers in order to increase commercial activity. In Chicago, engineers actually reversed the flow of the Chicago River to facilitate industrial growth. Cities also tend to pollute water systems, thus reducing the amount of fresh water and in some case posing major health hazards.

To theorize the city as an ecological unit is to open new possibilities for understanding the environmental inputs necessary for urban growth and the environmental impacts of urban growth. We not only have a large and increasing body of literature on the city as ecosystem, we also have some long-term study sites. In the US, for example, there are two urban ecological sites in Baltimore and Phoenix that have provided an array of interesting material for the modeling of urban ecosystems.[5]

Nature and the city

Nature is present in cities in often unforeseen and unplanned ways. Wildlife in a variety of forms continues to find ecological niches in the city. Urban tensions can be written through the narration of the relationship between cities and wildlife. Urban animal geographies can tell us a great deal about the city–nature dialectic, whether it is in the stories of rats in cities or the story of hawks in the city. Consider the case of Pale Male and Lola, two red-tailed hawks that made their nest in the façade of an exclusive high-rise apartment block in New York City's Fifth Avenue. Hawks have been noticed in the area since 1998 and every year the birds return to nest, breed and feed their young. Birdwatchers followed their progress through binoculars, cameras and websites. There is something heroic about the capacity of hawks to thrive in the city. Some residents of the apartment block thought otherwise. The president of the co-op board, wealthy real estate developer Richard Cohen, unilaterally ordered the nest removed in December 2004. Red-tailed hawks are rare enough to have been protected by a treaty signed in 1918 between several nations, including the US, Canada and Russia. An earlier attempt to evict the birds was blocked when their defenders invoked this international agreement.

However, the nest could be removed if it contained no eggs or chicks. The co-op board used this loophole and their decision initiated a major protest. Protesters dressed as birds mounted a vigil across the street from the building. The media publicized the story. One subtext was resistance to the power of the wealthy. Apartments at 927 Fifth Avenue can sell for as much as $18 million, and residents include the wealthy and the famous. The image of very rich residents evicting hawks from their perch was too delicious to ignore. The extremely negative publicity for the apartment building and its residents eventually led to a reversal of the eviction. Pale Male and Lola still nest on the building. You can follow the urban saga of Pale Male and Lola at their own website, www.palemale.com

Not all wildlife that shares the urban environment is as welcome as Pale Male and Lola. Rats for example have managed to find a home in most cities. They inhabit the dark tunnels and the hidden recesses of the city and have become symbols of disease and decay; they bring out fear and loathing rather than love and respect. And yet, like the cuddlier animals or the more photogenic birds, they too are urban survivors.

There is also a more self-conscious referencing of nature in cities. Consider urban parks. It is difficult to imagine London without Hyde Park, New York without Central Park or Washington, DC without the National Mall. Landscape architects such as Frederick Law Olmsted have left a permanent legacy in cities. The modern park movement is more closely tied into active participation than the environmental contemplation so beloved of the early park movement. City parks are now developed as much for their recreational opportunities as their aesthetic appeal. Urban planners realize that the successful referencing of nature is an important element in creating the right atmosphere, and it is often linked with the promise of economic redevelopment. Whether it is in the beaches of southern California, the lakeside shore of Chicago, or the parks of London and Paris, a commonly accepted attractive feature of urban life is the successful (re)incorporation of nature into the urban lifestyle, the city's image and the metropolitan experience.

An early urban ecologist

In 1864, the naturalist and geographer George Perkins Marsh in his book *Man and Nature*, commented that the likely cause of the decay of once flourishing civilizations such as the Roman Empire was in part due to acts of environmental neglect. He suggested, for example, that

Continued

Rome imposed on the products of agricultural labor in the rural districts taxes which the sale of the entire harvest would scarcely discharge; she drained them of their population by military conscription; she impoverished the peasantry by force and unpaid labor on public works; she hampered industry and internal commerce by absurd restrictions and unwise regulations. Hence, large tracts of land were left uncultivated, or altogether deserted, and exposed to all the destructive forces which act with such energy on the surface of the earth when it is deprived of those protections by which nature originally guarded it.

Source: Marsh, G. P. (1864) *Man and Nature: Or Physical Geography as Modified by Human Action*, New York: Scribner, pp. 10–12.

Urban sustainability

A central notion is urban sustainability; the idea that cities can be environmentally sustainable over the longer term. There is a widely shared belief that many cities impose such heavy environmental costs that the long-term future of the city may be undermined. The heavy reliance on fossil fuel, for example, and the increased use of the private auto as the main urban transport mode tend to degrade the environment. These problems are particularly acute in rapidly growing cities and cities in poor countries where environmental regulation is weaker and environmental improvement may not be considered an important political issue.[6] In China, for example, the rapid and often unregulated economic growth of the past twenty years has been purchased with the severe degradation of air, water and land. We can picture a three-stage model of the relationship between cities and environmental sensitivity; in the early stages urban growth is small and environmental impacts while strong are highly localized. As the city moves into a more industrialized mode of production, environmental degradation is more severe as the environmental impacts are heavier and longer lasting. As the economy matures and people become more affluent, a greater premium is placed on the quality of the urban environment. Environmental reforms are often instituted. Chinese cities are at this second stage of development. In cities around the world – rich and poor, developed and less developed – the struggle to live in a better urban environment, with clean water, fresh air and pleasant conditions is an important source of mobilization and platform for action.[7]

Questions of urban sustainability and environmental quality are intimately connected to issues of social justice; the worst environmental conditions are imposed on the lower-income most marginal urban residents. Poverty and environmental degradation tend to go hand in hand in a web of multiple deprivation and social exclusion. Houghton presents five principles of sustainable development that are also based on social justice: generation equity, intergenerational equity, geographic equity, procedural equity and interspecies equity.[8] He suggests approaches to achieving sustained development that include creating more self-reliant cities that reduce the environmental impact on the wider bioregion and redesigning cities so that land is used more effectively and rationally. Hough also lays out a road map for urban sustainability that recognizes the importance of maintaining the integrity of urban ecosystems.[9] One oft-touted example of a more suitable city development is Curitiba Brazil. This city of 1.6 million people developed a master plan in 1965 that limited central city growth and guided development to two north–south running corridors. The concentrated growth allowed the more effective use of public transport. There are now 1,100 buses that carry 1.4 million passengers a day and a network of pedestrian routes that allow people to travel by foot in the central business area. The net result is less demand for private car usage resulting in less pollution and a more pleasant urban environment. Many cities in the developed world can learn from Curitiba. Sustainable development practices can flow from poor to rich countries as well as from rich to poor countries. In many poor countries the need to husband scarce resources, recycle goods and reimagine the city provides a rich context for new urban practices.

A new approach

In the social sciences there is now an emerging body of literature that considers the environmental context of urban life and in the physical sciences a growing awareness that cities are environments worthy of serious ecological analysis. Let us end this introductory chapter with some brief examples of such recent work.

On the one hand, social scientists are examining the social context of the city–nature dialectic. Matthew Gandy's (2002) book *Concrete and Clay: Reworking Nature in New York* looks at the urbanization of nature in New York City and explores a series of relationships between nature, the city and social power in his consideration of the creation of the city's water supply, Central Park, the construction of urban parkways, a radical Puerto Rican environmental group in the 1960s and 1970s, and an anti-waste campaign in the Green Point–Williamsburg district of Brooklyn. He examines the environmental justice movement in a city where toxic facilities and land uses are consistently concentrated in minority-dominated areas of the city.[10] Erik Swyngedouw's (2004) book *Social Power and the Urbanization of Water* focuses on the city of Guayaquil in Ecuador, where 600,000 people lack easy access to potable water.

He shows that flows of water are deeply bound up with flows of power and influence, and water provision is not simply about connecting supply and demand but about the interconnections between the physical and the social, the environmental and the political.[11]

On the other hand some ecologists are using their techniques and approaches to consider the city as an ecosystem. Mary Cadenasso and colleagues have developed a model of the urban ecosystem as a complex of biophysical, social and built components. Using both watersheds and patch dynamics they seek to model the fluxes of energy, matter, population and capital with the goal of identifying the feedback between ecological information and environmental quality.[12] In a more detailed use of this urban ecology approach, Eric Keys and colleagues looked at the spatial structure of land use in Phoenix, Arizona from 1970 to 2000. They show how there was a marked change from agricultural to urban land and that remaining areas of desert were increasingly fragmented with implications for the urban ecology and biodiversity. They suggest the land use changes had an impact on higher levels of carbon emissions, creating a hotter urban heat island and a decline in native plant species.[13]

There is now a fascinating and important area of work in the space between the convergence of social scientists considering the nature of cities and ecologists looking at the city. This book will introduce the reader to this literature to show how the city is part of the environment, subject to environmental constraints and opportunities, a shaper as well as a container of environmental processes. The book is written from the assumption that we can only improve our understanding of the physical environment by considering its many and subtle links with the city and we can only enhance our social understanding of the city by exploring its many relationships to the physical world.

The subsequent chapters are organized into three parts that connect three central themes

- urban environments in historical context;
- urban environmental issues and;
- (re)aligning urban-nature relations.

The urban environment in history

The historical trend has been for urbanization to increase in size and scope, particularly during the twentieth century. Today half of the world's population now lives in cities and the size of cities has increased.

The emergence of cities has fundamentally altered the relationship between nature and society. In Chapter 2 we examine how early cities modified the physical environment while the physical environment also impacted cities. As cities emerged and grew, new environmental problems such as pollution and disease prompted new systems of regulation and infrastructural modifications that in turn modified the city-environmental relationship.

A major theme in the history of city–nature relations is the rise of the industrial city, in Europe and North America in the nineteenth and early twentieth centuries, and in the developing world since the latter half of the twentieth century. Chapter 3 notes how the rise of the industrial city generated vastly increased amounts of air, land and water contamination, but also brought about subsequent policy reforms and new forms of urban design. Coping with the problems of the urban environments during industrialization has been an important stimulus of new knowledge and the creation of new policies and infrastructure.

In Chapter 4 we take the story up to the present day and introduce five trends in the current wave of urban transformation; the impact of giant urban regions and megacities; postindustrial cities and brownfields, urban sprawl, new industrial spaces and shantytowns. Many cities have more than one trend occurring simultaneously.

Urban environmental issues

In the second part of the book we look at contemporary urban environmental issues from a variety of perspectives. We note that urban studies has focused on *location* rather than *site* and show in Chapter 5 how the occupancy of specific sites – such as deserts, beaches and flood plains – creates constraints and opportunities and the setting for the production of urban images and the urban–nature dynamic.

Since 9/11 and the Tsunami of 2004 and Hurricane Katrina in 2005, it has become increasingly clear that cities are vulnerable to disasters. In Chapter 6 we explore this vulnerability and argue that there is no such thing as a "natural disaster". We prefer the term environmental hazard/disaster because it highlights the social, economic and political forces that mediate or exacerbate hazards. At the same time we acknowledge that the city is also resilient. Efforts to recover and rebuild after fires and earthquakes, for example, often provide the impetus for urban growth. A discussion of urban hazards and disasters provides an important facet of the social-nature dialectic.

In Chapter 7 we document the emergence of urban ecology. Looking at cities as ecological systems holds out enormous promise for combining the insights of traditional ecology with the perspective of critical social science. Developments in this field of study allow us to look at cities more precisely as social–biophysical complexes.

We also consider issues of pollution, another angle on understanding urban environmental issues. In Chapters 8, 9 and 10 we examine water, air and garbage issues, respectively. In some cases cities have made significant improvements and advances. For many megacities in the developing world it is a mixed picture with some positive developments but there are also instances of cities being almost overwhelmed by the sheer size of growth in relation to the available resources.

(Re)aligning urban–nature relations.

The third part of this book considers theoretical and practical issues of realigning urban–nature relations. In Chapter 11 we investigate how issues of class, race and gender interconnect with urban environmental issues.

There are emerging discourses that attempt to redesign, recreate and rethink cities within a larger framework of livability and sustainability. In Chapter 12 we consider Slow Growth and New Urbanist movements, Smart Growth imperatives and historic preservation as examples of these new urban environmental discourses. We also discuss urban sustainable development as a major theoretical and practical discourse that seeks to reconnect cities to their local, regional and global ecologies.

The impetus for this book is our belief that environmental issues are increasingly urban based and that environmental issues are not incidental to the urban condition, they are central.

Guide to Further Reading

Benton, L. M. and Short, J. R. (1999) *Environmental Discourse and Practice*, Oxford: Blackwell.

Boone, C. and Moddares, A. (2006) *City and Environment*, Philadelphia: Temple University Press.

Breuste, J., Feldmann, H. and Ohlmann, O. (eds) (1998) *Urban Ecology*, New York: Springer-Verlag Telos.

Collins, J. P., Kinzing, A., Grimm, N., Fagan, W., Hope, D., Wu, J. and Borer, E. T. (2002) "A new urban ecology", *American Scientist* 88: 416–425.

Girardet, H. (2004) *Cities People Planet: Liveable Cities for a Sustainable World*, Chichester; Wiley.

Heynen, M., Kaika, M. and Swyngedouw, E. (eds) (2006) *In The Nature of Cities*, New York and London: Routledge.

Hough, M. (2004) *Cities and Natural Processes*, London: Routledge.

Keil, R. (2003) "Urban political ecology", *Urban Geography* 24: 723–738.

Ravetz, J. C., Roberts, George, P. and Howe, J. (2004) *Environment and the City*, London: Routledge.

Robbins, P. (2004) *Political Ecology: A Critical Introduction*, Oxford: Blackwell.

Short, J. R. (2005, orig 1991) *Imagined Country: Environment, Culture and Society*, Syracuse: Syracuse University Press.

Whatmore, S. (2002) *Hybrid Geographies: Natures, Cultures, Spaces*, London, Thousand Oaks, CA: Sage Publications.

2 Environmental issues in cities – a brief history

The first cities

Although there are on-going debates about exactly when, where and why the first cities arose, most scholars argue that the first, ancient cities began to emerge some five or six thousand years ago in various regions around the world. They began first as a shift from tribal communities and villages to larger, more complex, social, economic and political systems. The earliest cities were found in Mesopotamia (cities such as Ur, Erech, Lagash and Larsa that flourished in the southern portion of the Tigris–Euphrates river valley areas), Egypt (along the Nile such as Heliopolis, Memphis and Nekheb) and the Indus Valley (Harappa and Mohenjo-daro). Scholars estimate that at its height Ur might have had a population of 25,000. In China, the Huang Ho Valley appears to have been the region where the first cities of Shang and Chengchow in eastern Asia emerge. Grecian cities such as Thebes and Troy began to emerge around 1200 BCE, while the city of Rome began as a cluster of villages along the Tiber a few hundred years later. The Mayan cities of Tikal and Uaxactun are among the oldest in the New World, dating back to 200 ACE.

The world's first cities appear to have arisen in regions where climate and soil allowed the land to provide an abundance of plant and animal life that would be necessary to support larger populations. However, it was not just an agricultural surplus that created cities but the implementation of social power that directed labor and the production of a surplus that in turn allowed the development of cities. Agricultural surplus did not create cities, cities created agricultural surplus.

The rise and fall of Mayan cities

The Maya lived in the area in Central America which now consists of Yucatan, Guatemala, Belize and southern Mexico. Mayan cities include Tikal and

Continued

Uaxactun, Chichen Itza, Mayapan, Copan and Palenque. Tikal's population is estimated to have been around 60,000, which would give it a population density several times greater than the average European city at the same period in history. The Maya were highly accomplished in astronomy with an intimate knowledge of the calendar. The cosmology of the Maya permeated their lives and structured their cities. Cities were designed to coincide with astronomical rhythms. At Chichen Itza, during sunset, a sun serpent rises up the side of the stairway of the pyramid called El Castillo on the days of the Spring and Autumn equinox. In Mayan sites in the central and southern lowlands many temples have doorways and other features that align to celestial events.

At the heart of the Mayan city existed large plazas surrounded by the most important government and religious buildings. The towering palaces and the emphasis on height and verticality give an imposing air and reflect the desire to reach the heavens as well as reinforce the notion of a rigidly vertical social hierarchy. The most important religious temples sat atop the Mayan pyramids, which were impressively decorated. One theory suggests that these temples might have served as propaganda since they were the only structures that could be seen from vast distances. Outside of the center were structures of lesser nobles or smaller temples. The act of building itself required tremendous manpower and the ability to control labor.

Beginning in the eighth century and continuing for some 150 years, the great Mayan cities were abandoned, as wars raged and people fled. By 930 ACE, the Mayan heartland had lost 95 percent of its population. This prolonged event, known as the "Mayan collapse", is one of the enduring mysteries of pre-Columbian America. Some speculated an invasion and ensuing war might have led to the decline, but some have cast their eye on environmental factors. In 2003, scientists suggested that a 200-year dry spell, starting around 750 ACE, caused widespread droughts and a significant decline in regional rainfall.

Sources: Coe, M. (2005, 7th edn), *The Maya,* New York: Thames and Hudson. Faust, B., Anderson, E.N. and Frazier, J. (eds) (2004) *Rights, Resources, Culture and Conservation in the land of the Maya,* Westport, Conn: Praeger. Haug, G., Gunther, D., Peterson, L., Sigman, D., Hughen, K. and Aeschlimann, B. (2003) "Climate change and the collapse of the Maya civilization", *Science* 14 (5613): 1731–1735.

Water was one of the most critical elements. Almost all cities were located along major rivers and based their power (and that of their rulers) on the control of irrigation systems that served the surrounding countryside. The urban historian Lewis Mumford notes that it is no accident that the first cities began in river valleys.[14] Water management was an important ingredient in the development of centralized power. Large-scale engineering projects were only possible with centralized planning and hierarchic authority.

Cities are, in essence, a transformation of the physical environment to a built environment. Thus the environmental impact of cities is multidirectional: the city transforms the surrounding area, affecting the natural environment. In turn, the natural environment provides critical natural resources, and can also impact the city.

The very construction of the early cities involved an environmental transformation. In Middle America, around 2,500 years ago, the Zapotec Indians began building a great city, possible the first in the New World. The task involved the reshaping of Monte Albán, a 1,500-foot hill overlooking the Valley of Oaxaca in central Mexico. Cutting into the hillsides, workers constructed hundreds of terraces, stepped platforms with retaining walls designed mainly for plain and fancy residences.[15] In order to create a massive center square, workers toiled to flatten out the entire top of the hill by some 200 to 400 meters; it was a major feat of engineering and eight times bigger than St Peter's Square at the Vatican.[16] Monte Albán endured for more than one thousand years, housing some 20,000 to 30,000 people at its height.

The early cities referenced nature within the city walls. In Uruk it was said half the city was dedicated to open spaces with a greenbelt of market gardens.[17] Nebuchadnezzar II (604–562 BCE) is credited with building the legendary Hanging Gardens in Babylon in Persia. The Greek historian Diodorus Siculus provides a descriptive account of the gardens:

> The approach to the Garden sloped like a hillside and the several parts of the structure rose from one another tier on tier … On all this, the earth had been piled … and was thickly planted with trees of every kind that, by their great size and other charm, gave pleasure to the beholder.

Another noted,

> The Hanging Garden has plants cultivated above ground level, and the roots of the trees are embedded in an upper terrace … Streams of water emerging from elevated sources flow down sloping channels … These water irrigate the whole garden saturating the roots of plants and keeping the whole area moist.

Hence the grass is permanently green and the leaves of trees grow firmly attached to supple branches … This is a work of art of royal luxury.[18]

While our million plus cities seem dauntingly large today, early cities also boasted large populations. Athens, at its most successful in 431 BCE contained 300,000 people, while five centuries later Rome boasted 650,000. Chang'an in China reached one million and Teotihuacán in Mexico reached 200,000. Some were small in physical size but densely populated. The great Buddhist city of Taxila situated at the foot of the Himalayas covered just a few hectares, while the harbor at Carthage was not much larger than a soccer field.[19] By the thirteenth century, Paris, Milan and Venice contained populations of at least 100,000; by the end of the sixteenth century, London had reached 250,000. All of these cities were challenged by a variety of urban environmental issues such as food and water supply, disease and poverty, traffic congestion, limited housing and energy supplies.

Humans transformed the environment by creating the earliest cities. In the rest of this chapter we will select just three of the many environmental impacts: urban design, disease, and pollution controls.

Urban design

We will deconstruct ways in which the physical design and construction of the city reflect larger environmental discourses. The layout and design of city spaces convey messages about how people view the natural world. Urban design and urban planning also connect to wider social issues of power: most urban design resulted from the desires and decisions of a few powerful individuals.

Urban land use patterns are ways in which urban society utilizes and defines their relationship to the physical environment. And while the physical patterns – the arrangement of streets and parks and piazzas – vary widely from Berlin to Calcutta to Constantinople, many cities share common land use differentiations that include the pre-eminence of the central areas, the marginality of the periphery, and the location of particular crafts and merchant activities. There are two land use patterns that have endured since the earliest cities: walls and grids. Beyond their function, both symbolize deeper perceptions about the urban environment.

Fortifications are found in almost any old city of significant size and importance. Such fortifications ranged from wooden palisades and stone walls to ditches and moats. Citadels, or forts, were often incorporated as part of the walled fortification as was the case in Copenhagen and St Petersburg. Many cities were walled for defense against the "outside" and people could enter or leave only by the gates.[20] The wall served as a military device, a way to protect a city's market privileges, and a way to control the urban population. Encircling moats and canals added to the defense of the city. But these defensive structures also symbolically provided a clear demarcation line between urban and wilderness, between the civilized and the savage, between the insider and the outsider, between people and wild animals. Contact with the world outside was focused at specific access points, often gates. Those cities without walls often ended ambiguously with a few straggling buildings, and then fields. City walls are important for they show an exercise of control over the physical environment, one that attempts to segregate space into distinctive spheres. Walls might have also served as status markers, reflecting the display of power by political, religious or economic elites. In many cities, walls dominated the visual landscape and were the first structures seen when outsiders approached the city from the countryside. Walls inherently convey the message that urban space is separate and apart from the countryside and the wilderness.

Another common feature shared by cities around the world is the imposition of the rectangular street grid on urban space. The grid plan dates from antiquity; some of the earliest planned cities were built using grids and it is by far the most common pattern found in a variety of political societies from absolutionist powers to monarchies to democratic societies. The grid is a simple, rational order of packing the land, setting streets at right angles to one another. As early as 2600 BCE, Harappa (in North India) and Mohenjo-daro (in Pakistan) were built with blocks divided by a grid of straight streets, laid out in perfect right angles, running north–south and east–west. In Egypt, cities like Giza also used a common orientation: a north–south axis from the royal palace and an east–west axis from the temple meeting at a central plaza. In Babylon the streets were wide and straight and intersected at right angles. Teotihuacán, near modern-day Mexico City, is one of the largest grid-plan sites in the Americas, covering some eight square miles.

The rise of the Roman Empire standardized the grid plan, which became a common tool of Roman city planning. The Roman grid is characterized by a nearly perfect orthogonal layout of streets, all crossing each other at right angles. Typically gates were set at the midpoint on each of the four sides of the rectangle. The grid system allows for easy navigation and better flowing traffic; however, we can also see that this type of design imposes order on what would otherwise be an organic, chaotic physical environment. The grid system allows an imposition of power that determines the shape of living and working spaces. And, in some measure, the grid system inherently denies the importance or the existence of topography. The grid is the triumph of geometry over geography.

Water distribution and aqueducts in Ancient Rome

Beginning in the fourth century BCE, the city of Rome began construction on a series of aqueducts and established plans for the public distribution and management of water. Although the River Tiber flows through the heart of Rome, as the city grew, it needed more water. The historian E.J. Owens notes "a good supply of water was rightly regarded as one of the essential commodities of the maintenance of urban life in the ancient world". The aqueducts were a system of bridges, arches, and ducts that carried water from a remote source through an enclosed conduit (sometimes running underground, or above ground on an arcade bridge). Along the lines were settling tanks to remove foreign matter or filters that would keep the water free from debris, a factor that actually safeguarded public health. Once in or near Rome, water from the aqueducts passed into large, covered catch basins. The catch basins then distributed water through free-flowing canals, lead pipes and terra cotta pipes to storage reservoirs and public fountains. One of the more interesting aspects of the water supply systems was that most Romans retrieved their domestic water from public fountains; few had private supplies. Indeed, the tradition of communal civic usage of public works in part shaped the delivery system of water in Rome. With increased urbanization and population growth and rising issues of public health, Roman engineers separated different aqueducts and arranged for some to be used for drinking, with other lines being assigned other functions, according to water quality. Water was provided for a variety of uses including fountains and latrines, public baths and street cleaning.

Roman engineers became renowned for their hydraulic technology in general and the construction of aqueducts in particular. Through the development of water management techniques and the engineering of structural forms to transport, divert and store water, populations were able to settle and develop areas of Rome such as the site of the Roman Forum, which was previously a waterlogged swamp. In some regards, the feats of Roman engineering and the construction techniques employed in the building of aqueducts and delivery systems mark the urban foundation of Rome just as much as the Capitol or the forum.

Sources: Evans, H. (1994) *Water Distribution in Ancient Rome: The Evidence of Frontinus*, Ann Arbor: University of Michigan Press. Hodge, A. T. (1992) *Roman Aqueducts and Water Supply*, London: Duckworth. Owens, E. J. (1991) "The Kremna aqueduct and water supply in Roman cities", *Greece and Rome* 38(1): 41–58.

Perhaps one of the most striking examples of the use of the grid system is found in the Forbidden City, located at the exact center of ancient Beijing. The Forbidden City is surrounded by a large area called the Imperial City. As the imperial palace for both the Ming and Qing dynasties (1368–1911 ACE), the Forbidden City covered an area of 72,000 square meters and contained more than 9,000 rooms. Figure 2.1 shows the map for the Forbidden City and Figure 2.2 shows a view from the center of the palace. Emperor Yongle of the Ming Dynasty began building the Forbidden City in 1406; it took one million laborers and 100,000 craftsmen 15 years to complete.

A few features are worth noting. The rigid, rectangular layout was purposefully aligned to the cardinal directions, north, south, east and west. The entire palace, rectangular in shape, is surrounded by walls 10 meters high and a moat 52 meters wide. The walls were constructed to withstand attacks by cannons and are thick and squat. The wall has four gates with towers above them: East Magnificent Gate, West Magnificent Gate Meridian Gate to the south, and Gate of Divine Prowess to the North. On each of the four corners stand four turret towers. The layout within the walls is similarly geometric: the city is divided into northern and southern parts. The southern parts served as the emperor's work area; the northern parts contained his living quarters. These structures were arranged along a central axis at the exact midpoint and are symmetrical on either side.

Three structures in the southern area include the Hall of Supreme Harmony, the Hall of Central Harmony and the Hall of Preserving Harmony. The Hall of Supreme Harmony was the site of ceremonies, including the emperor's coronation, his marriage and other official state events. The Hall of Central Harmony was where the emperor rested and received officials; the Hall of Preserving Harmony was the site of banquets.

In addition to the use of the grid and the wall, the Forbidden City is replete with symbolic meaning that reflects Chinese culture, thus connecting to wider conceptions about the cosmos. The city itself was patterned after the Heavenly Palace. In ancient Chinese astrology, the Heavenly Palace centered around the North Star and was considered the center of heaven. The number nine received special inclusion in the city design. The number of houses in the Forbidden City is 9,999; and on every door are patterns of nine nails in vertical and horizontal lines.

The gridiron has prevailed in cities around the world – from Europe to Asia to the Americas. In US cities, the grid plan was nearly universal in the construction of new towns and cities. In some cases the grid softened over time. Boston, set on a grid in the mid-1600s, becomes more organic and curvilinear as it approaches the harbor.

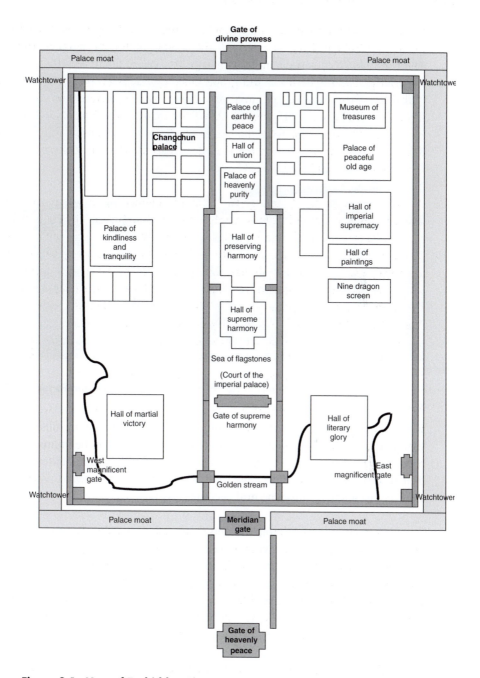

Figure 2.1 _Map of Forbidden City_

Figure 2.2 *Facing Tai he Dian (The Hall of Suprime Harmony) in the Forbidden City*
Source: Photo by Michele A. Judd

In San Francisco a grid was imposed on a dramatic topography of hills and valleys. Figure 2.3 shows Savannah, Georgia in 1734. Note that the entire area has been cleared of trees and vegetation: the natural world has been obliterated and the city (re)constructed from a *tabla rasa*. As US cities grew outward, particularly after the twentieth century, the grid becomes less prevalent.

Despite a multitude of geographies and topographies, of altitudes and latitudes, many cities on the grid share common design features: a lack of sensitivity to the physical environment, the imposition of the grid regardless of topography, a focus on the geometric (geometry over geography) and an underlying sense of the control of people-made space.

Most cities have gardens, small parks or large parks and other "green" amenities. These can also tell us about how society viewed the natural world. A garden is an arrangement of nature, whose plant materials and ordering principles are determined by prevailing ideas about the relationship between nature and society.

In Florence, Italy, the Boboli Gardens were constructed on a grand scale during the Italian Renaissance. Part of Renaissance culture saw the universe as a series

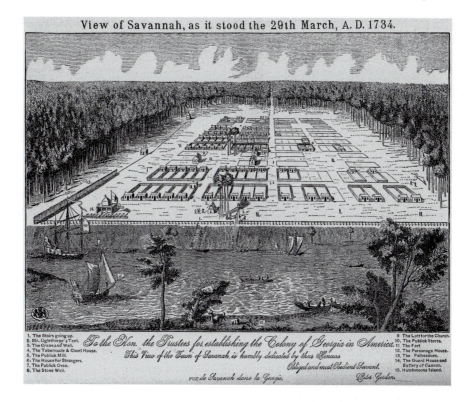

View of Savannah, as it stood the 29th March, A. D. 1734.

1, The Stairs going up. 2, Mr. Oglethorpe's Tent. 3, The Crane and Well. 4, The Tabernacle & Court House. 5, The Publick Mill. 6, The House for Strangers. 7, The Publick Oven. 8, The Stone Well. 9 The Lott for the Church. 10, The Publick Stores. 11, The Fort. 12, The Parsonage House. 13, The Pallisadoes. 14, The Guard House and Battery of Cannon. 15, Hutchinsons Island.

Figure 2.3 *Savannah, Georgia in 1734*
Source: "View of Savannah, as it stood the 29th March, A.D. 1734." From
***Report on the Social Statistics of Cities,** compiled by George E. Waring, Jr,*
***United States. Census Office, Part II, 1886.** Courtesy of the University of Texas*
Libraries, The University of Texas at Austin. Accessed at:
http://www.lib.utexas.edu/maps/historical/savannah_1734.jpg

of hierarchies – with God at the top, humans at the center and nature at the bottom. This larger metanarrative, or environmental discourse, thus informed urban design. Wealthy Italians referred to the gardens of their private residences as "villa gardens". "Villa" meant that all the formal parts of the ground were arranged in direct relation to the house; they were seen as extensions to the palaces and were treated as living space. The historian Claudia Lazzaro, in her book *The Italian Renaissance Garden*, notes that "Gardens in Renaissance Italy are witness to the attitude of contemporaries toward nature, much of which was inspired by classical culture."[21] The goal of many Renaissance garden designers was to implicitly recreate the gardens of classical antiquity, which were characterized by planting trees in ordered rows, clipping boxwoods into ornamental animals and shapes, decorating grottoes and placing statues of the gods of pagan mythology around the gardens. In this way, many Renaissance gardens were as much "art" as they were nature; as much contrived and manipulated as they were organic. Many gardens required significant earth-moving and water-powered devices. The gardens were dominated by this formal approach to design, although some aspects of the view that nature was "uncontrollable" also informed design. For example, many small bed areas were allowed to grow at will; groves of trees were not always planted in orderly rows.

The Boboli Gardens are a sixteenth-century Medici garden, located on the grounds behind the Palazzo Pitti, which was purchased by Eleanor de Toledo, wife of Cosimo I Medici, in 1549. The Medici's wealth originated from the wool trade; later they became influential international bankers. Politically, the Medicis were among the most powerful families in Florence during the Renaissance. Eleanor was instrumental in the creation of the gardens, hiring Niccolò Pericoli, known as Tribolo, a famous architect, to design the gardens between 1550 and 1558. Tribolo's plan was to center the gardens on a large fountain, framed by vegetation. But the plan had to incorporate a U-shaped hillside behind the palace, which Tribolo initially planned to plant with a 'boschetti' of evergreen trees planted in rows. Dwarf fruit trees were planted in large beds and near the orchards was a fishpond. There was also an extensive botanical garden.

Tribolo's plan was the basis for all the royal gardens in Europe, including Versailles. The Palace and Grounds of Versailles remain one of the most famous gardens in the world, although the term "garden" is an understatement. Similar to the Boboli Garden, Versailles was as much a political statement about power and the ability to display this in monumental scale, as it was about engaging with nature. The park and garden were designed by André Le Nôtre between 1661 and 1700. His plan included magnificent features, parterres, great basins, an orangery and even a canal. Avenues project from the palace toward different horizons, bringing together forest, garden, palace and city. The most visually obvious design element is the use of geometry – circles, diagonals, squares, rectangles both formalize and define different planting beds and spaces which were immaculately clipped and maintained (see Figure 2.4). A vast collection of outdoor sculpture and fountains add to the sense of grandeur. Importantly, the gardens of Versailles bespoke the power of an aristocracy; this was not a public garden in the twenty-first century definition.

The design of Versailles is intricately linked to wider intellectual developments of the time. The historian Carolyn Merchant has argued that the rise of the Scientific Revolution of the sixteenth and seventeen centuries was a time when the cosmos ceased to be viewed as an organism and became instead a machine.[22] She argues this was a dramatic shift in a broader world view of nature–society relationships.

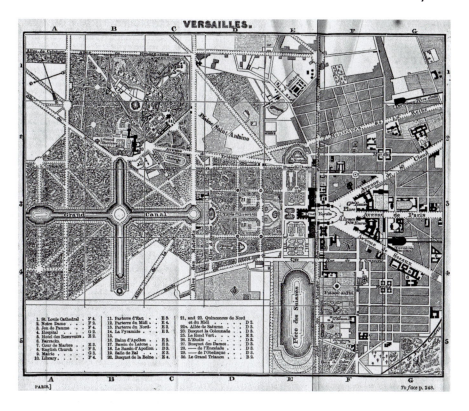

Figure 2.4 Versailles
Source: This image comes from the 4th edition of Meyers Konversationslexikon published in 1885

Advancements in physics, mathematics and technology impacted politics, literature, art, philosophy, religion and, we argue, even urban design. This crucial intellectual shift, which Merchant calls "the death of nature" is characterized by accelerated exploitation of human and natural resources in the name of culture and progress.[23] This new world view saw the earth as a machine, not only sanctioned for exploitation but also subjugation. We see this reflected in the plans for Versailles in which the defining design elements are rationality and symmetry, revealing an orderly, subdued nature, contained within pots and beds, shaped and clipped to repress the organic growth patterns. Nature had become an object to be observed, manipulated and ordered. Even statuary reminded the viewer of human handiwork, not nature's. While this era produced many of the famous gardens of Europe, we must also see these as products of aristocratic government and broader statements about changing views toward nature, particularly as society appears to be reducing the mysteries of the cosmos and gaining a sense of control over the natural world.

William Penn's green city

William Penn, the founder of Pennsylvania, designed the city of Philadelphia, one of the earliest attempts at utopian city planning. Penn, an ardent proponent of Quakerism and liberal government, was idealistic and also a member of the landed gentry with control over how land was dispersed. After King Charles II granted him 45,000 acres of land, Penn set aside 10,000 acres for Philadelphia. Penn chose the site at the convergence of the Delaware and Schuykill rivers where a safe, deep harbor provided good access to the interior of the state. He first advertised the layout of the city in 1682. Like most early US cities, it was designed around the rectangular gridiron and the concept of uniform streets and symmetrically placed houses. Two of the broadest avenues, Broad and High, crossed each other at Centre Square, and divided the city into four quadrants. Each quadrant contained green space in the form of a small park (now Washington, Franklin and Rittenhouse Squares and Logan Circle). These small parks were to be open to all. Despite the two riverfronts, Penn's city had an inward-facing design, focused on the central plaza.

It is possible that Penn was reacting against London, a city characterized by no discernible street pattern, housing that was built too close together, a lack of open or green spaces, and a physical structure that made its residents more vulnerable to the spread of disease or destruction by fire. In contrast, Penn's Philadelphia had an orderly geometric pattern of a rectangular grid divided into quadrants, public parks and substantial lot sizes for gardens. Lots of either one acre or half an acre were evenly spread across the width of the city, with the prime lots facing the Delaware and the Schuykill. The one acre or half an acre lots were large enough for each house to have a garden, in stark contrast to the more cramped cities of Europe. Penn's plan is remarkable for its spaced regularity and also for the concept of a "greenbelt" encircling the metropolis. While Philadelphia eventually outgrew Penn's original plan, his legacy would continue to influence local and national urban planning. His use of wide streets, the reservation of small garden parks, foreshadowed many future urban planning innovations.

Sources: Reps, J. (1965) *The Making of Urban America: A History of City Planning in the United States*, Princeton, NJ: Princeton University Press. Soderlund, J. (1983) *William Penn and the Founding of Pennsylvania, 1680–1684: a documentary history*, Philadelphia: University of Pennsylvania Press.

The development of another political capital, Washington, DC, underscores the attempt to embody political ideals with a developing sensitivity to the natural world. The 1791 design of the new federal capital reflected the struggle to create a space representative of American ideals and aspirations. The city was from the beginning rich in the physical and symbolic expressions of democratic ideals and a reflection of the formulation of national identity.

Pierre Charles L'Enfant was born in France and trained as an artist and painter under his father at the Royal Academy of Painting and Sculpture in Paris from 1771–1776. In 1777 at the age of 23, L'Enfant came to America to volunteer in the Continental Army. L'Enfant rose to the rank of major by the end of the Revolutionary War. George Washington knew of L'Enfant's artistic skills from the sketches and portraits of officers he had done while at Valley Forge.[24] After the war, L'Enfant established himself as an architect and worked in New York and Philadelphia. President Washington asked L'Enfant to design the new federal city.

L'Enfant arrived at Georgetown in March of 1791 intent on designing a capital that would befit a great new country. The concept of a planned city was not new in America. European and American examples were well known to Washington, Jefferson and many other of L'Enfant's contemporaries. L'Enfant himself knew of the plans for the American cities of Annapolis, Savannah, Williamsburg, Philadelphia and New York. He was also well versed in European city planning.

The neoclassical spatial order of the capital city was partly inspired by monumental planning trends in France. Common in France was the use of radial patterns imposed upon orthogonal streets, which offered an urban typology for expedient circulation, enchanting entry vistas and possibilities (space) for defining neighborhoods.[25] Within these monumental plans were symbolic spaces (from which emanated the radial streets). Designing celebratory public spaces in squares and semicircles was supposed to give the city an air of grandeur, delight and edification. All of these planning elements were well known to L'Enfant and eventually incorporated into his design.

In 1791, L'Enfant drafted a comprehensive plan for the new city of Washington, DC. The plan consisted of both a large map and a series of descriptions (Figure 2.5). L'Enfant's broad vision of a capital of buildings, public squares and promenades reflected the new country's optimistic outlook.[26] It was a design intent on celebrating a ceremonial city, the center of national government and culture. The plan was on an immense scale, far beyond the size or even expectation of the government at the time.[27]

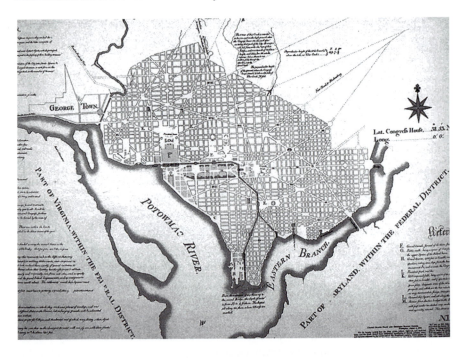

Figure 2.5 Map of Washington, DC 1791
Source: McMillan Commission Report, 1901. Image courtesy of the National Coalition to save Our Mall, http://www.savethemall.org/mall/resource-hist02.html

L'Enfant's plan is based on a template of a grid system of streets on a square block pattern, within which broad diagonal avenues would link the main hills. The resulting circles and squares provided public "reservations" or public space throughout the city. The diagonal avenues were named after the states. Central to the plan were two hills—one called Jenkins Hill, which would be the site of Congress and the other to the west would be the President's Palace (later called the White House). L'Enfant intended the two critical government buildings to be separated by space, a metaphor for the need to separate the branches of government. The broad diagonal street connecting the two was designated Pennsylvania Avenue. The two elevations were to be linked also by two large parks. One would stretch south from the President's Palace, and from the Congress House westward would be another he termed the Grand Avenue (eventually called the National Mall).[28]

Thus a central design element was the axial alignment of the city. L'Enfant envisioned a primary axis east-to-west stretching from the Capitol to the banks of the Potomac. The secondary axis going north-to-south would be perpendicular, crossing south from the President's House to the Potomac River. Where the two axes crossed, the convergence point, L'Enfant intended to place the statue of President Washington on horseback.

A unique feature of the plan was the inclusion of an intentionally large green space. The National Mall was the centerpiece of the plan. L'Enfant referred to it in his notes as a

> Grand Avenue, 400 feet in breadth and about a mile in length, bordered with gardens, and ending in a slope from the houses [of diplomats] on each side. This Avenue leads to Monument A [an equestrian figure of George Washington], and connects the Congress garden with the President's park…

It would be composed of a tree-lined walkway perhaps planted in the natural, picturesque style of landscape gardening gaining popularity in France and Britain or the more formal landscaping that was common in European cities.[29] He noted that the stretch of land would remain "a vast esplanade" with the "sort of places as may be attractive to the learned and afford diversion to the idle". His proposal that the major axis of the new city should be two great parks meeting at a central point (the Washington statue) was unusual in that most cities were built around commercial streets.

L'Enfant's plan for the Mall reflected his vision that a democracy should have public open space. In addition, the placement of the Mall as the centerpiece of the design represents the idea that open space conferred visible power and strength and symbolized an open, not closed, society. These were important values in the fledgling republic, which sought to distinguish itself from the British monarchy. It was intended that the Presidency be viewed physically and symbolically as "open" and accountable to the public. L'Enfant's plan embodied the history of the founding and early organization of the federal government, paying attention to the balance of power. But L'Enfant also included green space – a vast open space – at the center of the city. His Washington plan predates by more than seventy years the urban public park movement.

Disease and cities

From their beginnings until the twentieth century, cities have been places of disease. The gardens of Babylon and the temples of Egypt were emblems of urban glory, but the alleys in their shadows were choked with garbage, vast amounts of human and animal wastes, and decaying food.[30] The city was an environment that promoted and disseminated a range of diseases. Each of the four major types of disease transmission – airborne, waterborne, direct contact, and by insects or other vectors – was enhanced by urban life.

The leap from nomadic existence to urban life was not without its consequences. As the historian Arno Karlan notes, the emergence of agriculture and the domestication of animals generated an epidemiological crisis.[31] For the first time, people lived in continual intimate contact with other species: inevitably they exchanged pathogens. Pathogens made their way to humans who breathed the same air and dust, who touched animal wastes or butchered their bodies, used their wool and hides and consumed their milk, eggs or flesh. Each new domesticated species – dogs, birds, pigs, goats, cows, chickens, cats and even rats – exposed human populations to viruses and bacteria that, over time, mutated and thus entered human populations. It is speculated that measles is related to the viruses causing canine distemper, influenza is from pigs and horses and smallpox is related to those causing vaccinia in cows. Typhoid, often fatal to humans, may have originated in rodents and birds. Other vector-borne diseases include those spread by rodents, birds, snails, but perhaps the most varied and numerous vectors are insects. Diseases such as malaria, yellow fever and dengue came with greater exposure to mosquitoes, due to human-induced changes in the landscape brought about by agriculture and settlement such as ditches and puddles of irrigated fields, which became ideal breeding grounds for mosquitoes. Portrayed by some as the pinnacle of human ingenuity and civilization, cities became a breeding ground, transmission point and laboratory for a toxic cocktail of infectious diseases. Cities were very dangerous places and until urban public health reforms in the late nineteenth century, living in cities was always a hazardous activity. So, at second glance, the so-called triumph of the rise of agriculture and the development of cities turns out to have unleashed a public health hazard.

The first epidemic of a waterborne disease was probably caused by an infected caveman relieving himself in waters upstream of his neighbors. Perhaps the entire clan was infected, or perhaps the clan fled from the "evil spirits" ravaging their camp. But as long as people lived in small groups, isolated from each other, such incidents were sporadic. Once people began clustering in cities, they shared communal water, handled unwashed food, stepped in excrement from horses, or came into contact with urine used for dyes, bleaches and antiseptics. As farmers and villagers moved into cities, germs lurking in animals, wastes, filth and scavengers were offered feasts, and countless people were sickened and killed by previously unknown epidemics: smallpox, measles, mumps, influenza, scarlet fever, typhus, bubonic plague and the common cold. As cities grew, they become the breeding ground for waterborne, insect-borne and skin-to-skin infectious diseases. Typhus was most common, but typhoid, plague, smallpox, cholera and dysentery often broke out.

Fragmented stories and accounts of urban epidemics are found in ancient writings of the Sumerians, Babylonians, Egyptians, Greeks, Romans, Indians and Chinese. With the emergence and growth of cities the element of population density became a crucial factor in the spread of disease, particularly those infections spread via human contact such as coughing, sneezing, or casual contact with chamber pots. Thus "crowd diseases" such as smallpox, typhus, leprosy and tuberculosis were a major consequence of urbanization.

Poor sanitation along with population density were among other factors that amplified the spread of disease. Water, too scarce and precious, was used only for drinking and cooking; people rarely bathed or consistently washed their clothes. Street life in medieval England was dangerous and uncomfortable:

> butcher and poulterers were by no means alone in their careless disposal of animal refuse; fishmongers and cooks and the ordinary households were all guilty. In Chester women carrying entrails of animals from the butchers carried them uncovered and threw them out near the gates, to the public nuisance…The private citizen was only too ready to dispose of dead dogs and cats by dropping them in the river or just over the town wall.[32]

These conditions facilitated the plague spread from rats to people in crowded urban areas and there are recorded instances in early urban history. An outbreak of plague (bubonic or perhaps measles) in Athens in 430 BC was said to have destroyed the city's "Periclean golden age", unraveling its military strengths and its civic and moral fabric. For the next thousand years, the plague was a periodic, recurring disease. The cycle of outbreaks in the fourteenth century would be known as The Black Death. The first of several waves of plague descended on Europe starting in 1347 and lasting through 1450. The Black Death consisted of both bubonic and pneumonic forms of the plague. Bubonic plague, caused by flea bites spread by lice on host rats, caused swelling of the lymph nodes (buboes) and fever; the pneumonic form (spread person to person) invaded the lungs and was more deadly. It has been estimated that nearly 25 percent of Europe's population, some 25 million people, died in the 1347–1450 epidemic. As the Black Death spread across the cities of Europe it created panic, death and despair. The impacts were profound. The Italian writer Boccaccio wrote of his experiences living in Florence during the Black Death (1347–1349) in the introduction to the *Decameron*:

> Whenever, fairest ladies, I pause to consider how compassionate you all are by nature, I invariably become aware that the present work will seem to you to possess an irksome and ponderous opening. For it carries at its head the painful memory of the deadly havoc wrought by the recent plague, which

brought so much heartache and misery to those who witnessed, or had experience of it. What more remains to be said, except that the cruelty of heaven (and possibly, in some measure, also that of man) was so immense and so devastating that between March and July of the year in question, what with the fury of pestilence and the fact that so many of the sick were inadequately cared for or abandoned in their hour of need because the healthy were too terrified to approach them, it is reliably thought that over a hundred thousand human lives were extinguished within the walls of the city Florence.[33]

The plague would inspire other writers such as Albert Camus's "The Plague". One of the more well-documented outbreaks of plague occurred in London between 1665 and 1666, which at that time had a population of some 500,000. During the height of the outbreak, some 7,000 people died each week. More than 100,000 people had died by its end. The city was nearly deserted. The historian Walter George Bell chronicled the great London Plague and noted that it retreated in 1666 primarily due to the Great Fire of London which burned more than four-fifths of the city.[34] At the time, London's buildings were built using timber. As a result of the Great Fire the plague disappeared, and in its wake new urban design regulations and laws would substantially improve sanitation and would help control future outbreaks of diseases. Bell notes one of the more interesting proclamations encouraged people to take care of the cats, in order to control the rat population. New building codes required new buildings to be constructed of brick and stone, which would be both fireproof and more sanitary. Finally, laws mandated wider and straighter streets to eliminate congestion and to prevent the quick spread of fire in the future. Reforms enacted primarily in response to the Great Fire would also result in better sanitation.

The plague was not the only infectious disease to ravage urban populations. In the nineteenth century, cholera became one of the world's first truly global diseases. Today we know that cholera is caused by ingesting water, food or any other material contaminated by the feces of a cholera victim. Contact with a contaminated chamber pot, soiled clothing or bedding can spread the disease. The onset of disease occurs within 12–48 hours of infection and is often characterized by extreme diarrhea, sharp muscular cramps, vomiting and fever, and – sometimes – death.

The disease spread along trade routes and into cities where it spread from port to port (sometimes the germ lived in contaminated kegs of water or in the excrement of infected victims, and was transmitted by travelers). In the early 1830s, cholera entered New York through infected ships. Quarantine regulations which sought to "contain" cities were ineffective. The disease hit worst where poor drainage and human contact came together: the crowded slums.

In 1854, Dr John Snow made the biggest contribution to solving the mystery of cholera. By plotting the residential locations of those infected with cholera, Snow demonstrated how the cases of cholera in central London were clustered around a single source of contaminated drinking water, a neighborhood water pump on Broad Street (Figure 2.6). The Broad Street hand pumps received untreated water from the Thames. Although Snow is most famous for his analysis of the Broad Street outbreak, his analysis confirmed a theory he had been developing over the preceding years: that contaminated water contained disease-causing organisms. Snow's map of cholera would detail the spatial pattern to the geography of the spread of disease and ultimately led to the cause of the disease.

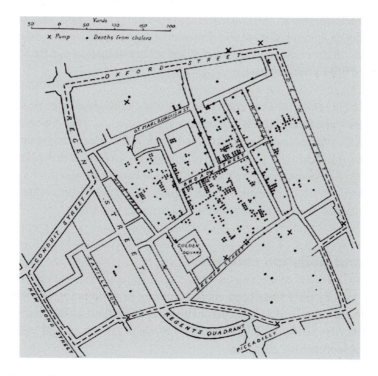

Figure 2.6 John Snow's map of cholera
Source: Drawn by Dr John Snow about 1854; shown in L.D. Stamp (1964)
A Geography of Life and Death

The connection between disease and cities is not simply a physical phenomena limited to epidemiological dimensions. Diseases are very much linked to social and political elements. For example, many early policies and approaches to sanitation characterized slums as places of crime; moral and social decay threatened disorder and disease.[35] The equation of immigrants, minorities and other slum dwellers with disease became a powerful image used in attempts to control the urban environment. In Victorian London, endemic poverty and disease and the potential violence of the factory laborer created anxiety amongst the Victorian elite who viewed these as a threat to progress and social order. The historian Maynard Swanson has argued that in the South African cities of Cape Town and Port Elizabeth, authorities used public fears of epidemic disease to justify residential racial segregation. Swanson noted that the fear of cholera, smallpox and plague rationalized efforts to remove or even segregate various elements of society. He concluded that "medical officers and other public authorities in South Africa were imbued with the imagery of infectious disease as a social metaphor, and that this metaphor powerfully interacted with British and South African racial attitudes to influence the policies and shape the institutions of segregation".[36] Outbreaks of diseases, and particularly an outbreak of bubonic plague in 1901, became an opportunity for those who were promoting segregationist solutions to social problems; thus concern for sanitation is one major factor in the creation of urban apartheid. Issues of disease and the fear of the other went hand in hand.

That cities survived the continued onslaught of diseases is remarkable. Because cities experienced both epidemics and endemics, they became "population sinks", where death rates exceeded birth rates. Many catastrophic epidemics were followed by a resurgence in population growth as the inflow of migrants from the countryside replenished the urban populations. As Arno Karlen observed, "it is a testimony to human vigor and adaptability that city life flourished despite plagues, famines, wars and migrations".[37]

Cities and pollution controls

As long as there have been cities, there has been pollution and attempts to deal with the issue. In 1231, Emperor Frederick II Hofenstaufen, concerned about air quality in Sicily, decreed a new law to clean up the air and went as far as to impose monetary penalties:

> We are disposed to preserve by our zealous solicitude, insofar as we are able, the salubrious air which divine judgment has provided. We therefore command that henceforth no one be permitted to place linen or hemp for

retting [soaking] in any waters within the distance of one mile from any city or castle, lest from this, as we have learned certainly happens, the quality of the air is corrupted. We order that the bodies of the dead, not placed in coffins, should be buried to a depth of one-half a rod. If anyone does the contrary, he shall pay our court one augustalis. We further order that those who take the skins of animals should put the carcasses and wastes which create an odor outside the territory [of a city] by a fourth part of a mile, or throw them into the sea or river. If anyone does the contrary, he shall pay to our court one augustalis for dogs and animals which are larger than dogs, and one-half an augustalis for smaller animals.[38]

At about the same time, the bailiff of the medieval city of York posted an order to deal with both air and water pollution in a city where the stench was pronounced:

Whereas it is sufficiently evident that the air is so corrupted and infected by the pigsties situated in the king's highways and in the lanes of that town and by the swine feeding and frequently wandering about ... and by dung and dunghills and many other foul things placed in the streets and lanes, that great repugnance overtakes the king's ministers staying in that town and also others dwelling and passing through, the advantage of more wholesome air is impeded, the state of men is grievously injured, and other unbearable inconveniences ... the king, being unwilling longer to tolerate such great and unbearable defects there, orders the bailiffs to cause the pigsties, aforesaid streets and lanes to be cleansed from all dung and to cause them to be kept thus cleansed hereafter.[39]

Throughout history there are many examples of such ordinances as local pollution became pronounced enough to force political action. However, we can identify two major shifts that prompted changes in national policies that impacted cities. The first, which we will explore in more detail in the next chapter, is the advent of the industrial city which prompted a variety of public health/pollution abatement regulations. The second is the environmental revolution of the mid to late twentieth century that created more stringent regulations to limit pollution and introduced national standards.

Take the case of air pollution, which had been increasing its impact with industrialization and population growth. Air pollution was a problem in British cities for more than 800 years. Until the twelfth century, most Londoners burned wood for fuel. But as the city grew and the forests shrank, wood became scarce. Large deposits of coal provided a cheap alternative. By the thirteenth century,

Londoners were burning soft, bituminous coal to heat their homes and fuel their factories. In 1257, Queen Eleanor of Provence visited Nottingham Castle and was forced to move because of fouled air, full of heavy coal smoke. Numerous attempts to control coal burning and to punish offenders were made during the thirteenth and fourteenth centuries but were largely ineffective. In 1661, John Evelyn wrote an anti-coal treatise *Fumifungium: or the Inconvenience of the Aer and Smoake of London Dissipated*, in which he pleaded with the King and Parliament to do something about the dark yellow mist with a pungent smell that hung over the city. By the middle of the nineteenth century London's air was so highly polluted that it figured in popular novels. Here is an extract from a Dickens novel *Bleak House*, first published in 1853:

> [A]s he handed me into a fly after superintending the removal of my boxes,
> I asked him whether there was a great fire anywhere? For the streets were so
> full of dense brown smoke that scarcely anything was to be seen.
> "Oh, dear no, miss," he said. "This is a London particular."
> I had never heard of such a thing.
> "A fog, miss," said the young gentleman.

Episodes such as Charles Dickens describes were fairly common in mid and late nineteenth-century London, a result of the burning of coal laden with high levels of sulfur released into the air when combusted. Throughout the fall months, during periods of calm, smoke particles from industrial plumes would mix with fog giving it a yellow-black color. If wind speeds were low at these times, this would cause the smog to stagnate, with higher levels of pollution close to the ground. Air pollution was long accepted as simply an unfortunate fact of London life. Increasing pollution does not lead automatically to environmental regulations; only when the problems reach high on the political agenda is action taken. In December 1952 an anticyclone settled over London. The wind dropped and a thick fog began to form. Londoners burned more coal to combat the winter cold. The Great London Smog, as it was named, darkened the streets of London for five days and levels of sulfur dioxide increased 7-fold and levels of smoke increased 3-fold. The smog killed approximately 4,000 people immediately and caused another 8,000 premature deaths. Most of the 4,000 deaths occurred due to pneumonia, bronchitis, tuberculosis or heart failure with the peak in the number of deaths coinciding with the peak in both smoke and sulfur dioxide pollution levels. The British government, initially reluctant to admit that coal smoke was the cause, blamed the deaths on a flu epidemic. But pressure mounted and in 1956 the government passed its first Clean Air Act, which created smokeless zones in the city where only clean burning fuels were allowed. The Clean Air Act differed from previous legislation because it controlled not only domestic air pollution, but also industrial. Within a few years after the legislation, smoke emissions from domestic sources declined by more than a third; and the city experienced an overall decrease of smoke concentrations by some 70 percent. Since the regulation, the reduction of sulfur dioxide has made London's infamous "peasoupers" a thing of the past.

The major shift in regulatory context came with the new environmental awareness. There were farseeing advocates and isolated pieces of legislation prior to this shift but afterwards environmental protection and pollution control became mainstream, national and indeed international concerns of major importance. Take the case of the US. Outbreaks of urban pollution in the US were the back-drop to the shift. In 1948, in Donora, Pennsylvania, between 20 and 40 people died from an episode of air pollution inversion. Dubbed the "smog tragedy", it focused attention on the growing problem of air pollution in urban areas. A sim-ilar smog disaster occurred in New York City in 1966 when 80 people died. In the 1950s, numerous beaches were closed due to water contamination and by the early 1960s the Animas River in New Mexico reported radioactive content in the water, while the Passaic River in New Jersey was so polluted that thousands of fish washed up on the shores. Cape Cod beaches were fouled by oil spills, as was the York River in Virginia. The year 1969 appeared to be the "year of disasters". Lake Erie was declared "dead", so polluted it was devoid of fish and aquatic life. Off the shores of Santa Barbara in California a large oil spill contaminated miles of shoreline, killing sea otters, birds and other marine animals while television crews filmed frantic volunteers trying to wash the oil off the dying animals. On June 22, railroad sparks set fire to the Cuyahoga River in Cleveland. The river, saturated with oil, kerosene, debris and other flammable chemicals was engulfed in a five-story-high blaze of flames. In Washington, DC, the Potomac River was clogged with blue-green algae blooms that were both a nuisance and a public health threat. In the nation's capital, the rivers were little more than open sewers. These and other episodes around the US served as a rallying point for the emergence of the modern environmental movement, and the passage of the 1969 National Environmental Policy Act. This watershed legislation established the Environmental Protection Agency (EPA) and charged the agency to study pollution and recommend new policies.

Conclusions

The emergence of cities altered the relationship between nature and society in pro-found ways. We have seen how the earliest cities modified the physical environment in the creation of specific urban designs and growing environmental impacts.

The physical environment also impacted the cities through long-term sustainability and as new incubators for disease. The case of pollution controls shows the example of a feedback system when growing urbanization created perceived environmental problems that in turn engendered new systems of regulation and infrastructural modifications that in turn involved changes to the city–environmental dynamic.

Guide to Further Reading

Bell, W. (1994) *The Great Plague of London*, London: Bracken Books.

Berg. S. (2007) *Grand Avenues: The Story of the French Visionary Who Designed Washington, D.C.*, New York: Pantheon.

Craddock, S. (2004) *City of Plagues: Disease, Poverty and Deviance in San Francisco*, Minneapolis: University of Minnesota Press.

Daunton, M. (ed.) (2001) *The Cambridge Urban History of Britain*, Cambridge: Cambridge University Press.

Hempel. S. (2007) *The Strange Case of the Broad Street Pump: John Snow and the Mystery of Cholera*, Berkeley: University of California Press.

Hope, V. (2000) *Death and Disease in the Ancient City*, London, and New York: Routledge.

Karlen, A. (1995) *Man and Microbes: Disease and Plagues In History and Modern Times*, New York: G.P. Putnam.

Mukerji, C. (1997) *Territorial Ambitions and the Gardens of Versailles*, Cambridge: Cambridge University Press.

Mumford, L. (1989) *The City in History*, San Diego and New York: Harvest Books.

Porter, Y. (2004) *Palaces and Gardens of Persia*, Paris: Flammarrion.

Reps, J. (1965) *The Making of Urban America: A History of City Planning in the United States*, Princeton, NJ: Princeton University Press.

Schott, D., Luckin, B. and Massard-Guilbaud, G. (eds) (2005) *Resources of the City: Contributions to an Environmental History of Modern Europe*, Hampshire, England and Burlington, VT: Aldershot.

Spary, E. (2000) *Utopia's Garden: French Natural History From Old Regime to Revolution*, Chicago: University of Chicago Press.

Spirn, A.W. (1984) *The Granite Garden: Urban Nature and Human Design*, New York: Basic Books.

An excellent website on John Snow is The John Snow Archive and Research Companion at matrix.msu.edu/~johnsnow/maps.php

3 The industrial city

The City is of Night, perchance of Death,
But certainly of Night; for never there
Can come the lucid mornings fragrant breath
After the dewy morning's cold grey air

James Thomson, *The City of Dreadful Night,* 1880

When James Thomson wrote *The City of Dreadful Night* in 1880, he was refer-
ring to the dirty, gritty city of London in the midst of rapid industrialization.
Thomson's London is beset by disease and doom. His image was accurate for
many of Europe's and America's industrializing cities in the eighteenth and nineteenth centuries. The onset of the industrial revolution profoundly and irrevoca-
bly shifted human relationships with their physical environment. And while it is
probably true that life in the cities was not necessarily worse than that in rural
areas, the problems of pollution and poverty and distress were more evident,
more massed, and less easy to ignore. For the most part, when cities were smaller
and density low, pollution was perceived more as a nuisance than a threat to
human health. The industrial era generated new agents in the city: the factory
and the railroad. Rapid urbanization and increasing population density created a
strained, hazardous and degraded physical environment that had visible and
often significant health impacts. But the industrial city was also the cauldron in
which new environmental-social relations were forged. New public health measures were introduced and the urban parks movement as well as the garden cities
movement were just some of the responses that reshaped, both in imagination
and practice, the urban–nature dynamic.

In this chapter, we focus on three important themes. The first is the increasing
and often visibly noticeable environmental contamination of water, land and air.
The second is the policy reforms of the industrial city that emerge as a response,
which continues to influence contemporary approaches to environmental reform.

The third is the new forms of urban design. We will draw our examples largely on the experience of the US and to a lesser extent the UK, but a similar tale can be told for most industrial cities around the world.

Pollution in Coketown

In 1844 Friedrich Engels published *The Condition of the Working Class in England*, based in part on his stay in Manchester in England. His description of the "irregular cramming together of dwellings", the streets of "filth and disgusting grime", the "coal-black, foul-smelling stream", and the general "stench of animal putrefaction" vividly conveyed his impression of the new urban form of the industrial city. He noted,

> If any one wishes to see in how little space a human being can move, how little air—and such air!—he can breath, how little of civilization he may share and yet live, it is only necessary to travel hither. Everything which here arouses horror and indignation is of recent origin, belongs to the industrial epoch.[40]

Later Charles Dickens, in his novel *Hard Times*, would coin the phrase "Coketown" to refer to cities like Manchester, a new type of city where industrialism produced the most degraded urban environment the world had yet seen.

The urban historian Lewis Mumford said Coketown was where "the immense productivity of the machine, the slag heaps and rubbish heaps reached mountainous proportions, while the human beings whose labor made these achievements possible were crippled and killed almost as fast as they would have been on a battlefield".[41] The economic foundations of the industrial city were the exploitation of the coal mine, the vastly increased production of iron and the use of steady, reliable mechanical power – the steam engine. These functions had a devastating effect on the physical environment.

The industrial city was characterized by mechanization and the intensification of the use of resources. Coal replaced wood as the leading source of energy. Mines were built to extract both coal and iron. New industries emerged such as smelting, and large-scale manufacturing. Coal, iron ore, lumber and petroleum were abundant raw materials. It was the age of steam engines, factories and smokestacks. A new working class of factory workers was created. The factory system encouraged the centralization of production in cities. Factories had to be near or have access to sources of raw material, a sufficient labor force and sizable markets.[42] But, as more and more factories concentrated in urban areas, the result was both increased urban population

Table 3.1 *Urbanization in Manchester, England*

Date	Estimated urban population
1685	6,000
1760	30,000
1801	72,275
1850	303,382

Source: Lewis Mumford (1989), *The City in History*, San Diego: Harvest Books, p. 455.

growth and the discharge of pollution in massive amounts. In Britain, people flocked to the new burgeoning industrial cities like Leeds, Halifax, Rochdale, Swansea, as well as Bristol, Liverpool and London where jobs flourished. Consider the rapid transformation of Manchester, England from a small mill city to a large industrial one as shown in Table 3.1.

Although the cities of Britain were at the forefront of the industrial revolution, the rest of Europe, and North America, were not far behind. Cities like Essen, Cologne, Toronto and Melbourne began to grow rapidly by the late nineteenth century. Swift economic change and population growth also occurred in cities such as Pittsburgh, Cleveland, Milwaukee, Boston and Philadelphia. Industrialization generated unprecedented levels of urbanization (see Figure 3.1).

Figure 3.1 *Industrial Pittsburgh, date unknown*
Source: Library of Congress, Historic American Buildings Survey/Historic American Landscapes Survey, call number: HAER, PA,650MONE, 1-

Industrial Pittsburgh and acid mine drainage

In addition to diminished drinking quality, acid also retards water delivery, as its corrosive action destroyed pipes and pumps. The cost of such damage was a key factor in the erection of neutralizing plants as a component of overall water treatment. The savings provided by neutralization offered a considerable return on the investment, not only at treatment plants, but also in the communities using the water, since the acidic water also destroyed domestic plumbing.

The impact on municipal water supplies is aptly illustrated by the experience of McKeesport, Pennsylvania. The city, located thirteen miles upstream from Pittsburgh at the junction of the Monongahela and Youghiogheny Rivers, was one of the region's most important steel production centers. McKeesport supplied its municipal water needs from wells and from the Youghiogheny. The river's drainage basin contained one of the most active coal-mining districts in the nation. The combination of large volumes of manufacturing wastes and domestic sewage mixed into high concentrations of acid drainage, produced, according to the United States Geological Survey, "the worst water in the United States". In a 1906 Water Supply and Irrigation Paper, the survey characterized McKeesport's water as "dangerous and in no sense potable or fit for consumption by human beings". Alternatives to municipal supplies included bottled and frequently contaminated well water. In 1903, McKeesport's typhoid-attributed mortality of 112 per 100,000 confirmed USGS condemnation.

In 1908, McKeesport put into operation one of the first municipal water treatment plants designed to filter particulates, kill bacteria through chlorination, neutralize acid and soften water. The plan treated ten million gallons daily. Total costs of pollution are difficult to estimate because damage is often subtle and not directly connected to the effluents. A 1910 pollution survey conducted by the Army Corps of Engineers estimated that acid drainage cost the Pittsburgh district $3 million a year. The Corps itself experienced $50,000 a year in damages to its river structures, primarily steel bridge supports. In a study conducted between 1914 and 1917, the United States Public Health Service suggested the total cost of damages from acid drainage may have been as high as $10 million a year.

Source: Casner, N. (2003) "Acid mine drainage and Pittsburgh's water quality", in J. Tarr (ed.) *Devastation and Renewal: An Environmental History of Pittsburgh and Its Region*, Pittsburgh: University of Pittsburgh Press. pp. 94-95.

The advent of the industrial economy and the trend toward specialized labor resulted in many trades and crafts that carried distinctive health risks. Laborers who skinned leather and worked with hides were exposed to anthrax all day, every day. Potters and miners were poisoned with toxic metals such as mercury, lead and arsenic.[43]

In Manchester, old, central, high density areas were associated with high mortality rates; in Liverpool, an 1871 survey showed that in streets where the death rate was high, a large proportion of deaths were of young children; similarly in New York City, the mortality rate for infants in 1810 was 120 per one thousand live births but rose to 180 per one thousand live births by 1850 and to 240 in 1870.[44] The causes were poor housing, lack of sanitation, lack of clean water, poor diet and endemic disease. Industrial cities experienced an environmental crisis on a scale not encountered previously. In short, the industrial city witnessed widespread environmental deterioration of water, land and air.

Water pollution in the industrial city

> *I wander through each dirty street*
> *Near where the dirty Thames does flow*
> *And mark in every face I meet*
> *Marks of weakness, marks of woe*
> William Blake

Water was a critical environmental element in production and manufacturing during the industrial revolution. It was an ingredient in many industrial processes and a convenient dumping ground for industrial byproducts. In the industrial era, cities faced two pressing water issues. The first was water quality. The second was locating sufficient water supplies for a rapidly urbanizing population.

Water pollution in the industrial city originates from two main sources. The first is residential: human and animal waste, which is composed of organic compounds. The second is commercial: factories, businesses. Commercial sources included both organic byproducts and, increasingly, inorganic byproducts as advances in technology helped to fabricate new inorganic materials such as plastics, dioxins and heavy metals. Factories, especially those involved in textile, chemical and iron and steel industries, were often sited by rivers or bays because they needed large quantities of water for production processes. Factories used water for steam boilers and to cool engines, and in the making and disposing of chemical solutions and dyes. Water, once used, was dumped back into rivers, creeks and tributaries. These resources became open sewers, poisoning aquatic life. By the 1870s, for example, the Passaic River in New Jersey was so polluted the city had to abandon it as a water supply and commercial fishing in the area was effectively ended.[45] In Chicago, the center of slaughtering and meat processing, nearly 1 million cattle and pigs were slaughtered each year. Meat plants routinely flushed carcasses and unwanted parts into the river, while glues, gums, dyes, margarine, fertilizer, sausage casings, brushes from processing and packaging plants nearby were also dumped into water sources. The Chicago River was so polluted that ice cut from the river released a disgusting stench as it melted.[46]

The poor condition of London's Thames made a great impression on poets and others. In one poem written in 1859, *State of the Thames*, the Thames is decried by a London satirist:

> *River, river, reeking river!*
> *Doomed to drudgery foule and vile;*
> *Noisome, noxious fumes distilling,*
> *Fumes which streets and housing filling*
> *Harpy life, defile*[47]

In Australia, the city once called "Marvelous Melbourne" was given the infamous title "Marvelous Smellbourne" for the stench caused by uncollected garbage, and its huge, open sewers and lack of water treatment. The River Yarra was as polluted as any urban European watercourse by the late nineteenth century.[48]

Human wastes were disposed of in cesspools and privies (although sometimes they were tossed out on the street or in vacant lots), hence most of the residential waste was put onto the land.[49] Before 1850 no US city had sewer systems for the collection of human waste or wastewater. In Europe some cities had sewer systems built of masonry dating back as early as the thirteenth or fourteenth century but many had fallen into disrepair. In Paris, for example, the sewers were decayed and poorly maintained and were considered dangerous places to go. In Victor Hugo's *Les Misérables*, the sewer was politically symbolic: sewers harbored enemies of the state, and outcasts such as thieves and prostitutes.

Prior to the nineteenth century, water supplies were obtained from mostly local sources such as wells, nearby ponds, streams and rivers. But for many industrial cities, by the 1860s population had outgrown water supplies. New York, Philadelphia and Boston were unable to provide an adequate supply of clean water to a growing population. This supply issue meant there was a shortage of clean water for drinking, bathing or cooking, but it also hampered efforts to fight fires.

The sewers of Paris

Paris above ground is an ever-changing panorama, which anyone can view by paying for it; sometimes the coin is simply money, or cheaper and better yet, a little enterprise or exercise; but too often it is a sight draft upon either health or morals ... Few, however, think of glancing at subterranean Paris; that mighty labyrinth of streets beneath ground, seen but rarely by human eyes, but without which Paris above ground would be an uninhabitable mass, or a generator of pestilence. There is nothing here for show, but all for use. Built to endure for ages, and to subserve the necessities of millions of human beings, performing in the material economy of social life functions as important and as indispensable as the veins and arteries in physical life, they are worthy of a glance, at all events, that we may learn the labor and expense in lighting, watering, and cleaning a modern capital. These indispensable offices are all moving quietly on in their prescribed paths, unseen and almost unknown by the millions of noisy feet above them. Yet, should any derangement ensue, the health and comfort of the city is at once in jeopardy. Were the Tuileries consumed by fire, and the Arc of Triumph engulfed in an earthquake, the Parisians would simply have two fine monuments the less. But were the drains, water, and gas of Paris to be suddenly arrested, the city would become uninhabitable, and the ancient marshes of Lutèce would regain their lost empire. It was not, however, until the commencement of the last century that a regular system of drainage was established ... the system has been continually improved upon, until it has rendered Paris the cleanest and best lighted capital in the world.

Source: "Life in Paris – sketches above and below ground", *Harper's Bazaar* February 1854: 306-307 retrieved at: http://www.sewerhistory.org/articles/whregion/1854_ah01/index.htm

Land pollution

Industrial cities generated new pollutants that were discarded on land. Factories sought the easiest methods of disposal for non-liquid wastes. Rubbish, garbage, ashes, scrap metals and slag (formed during iron smelting and other metallurgical processes) were often disposed of on open or vacant lots around the city.

Many industrial cities were ill-prepared to provide adequate sanitation services. In Paris, the basic rule for dealing with garbage was "*tout-à-la-rue*", translating as "all in the street", including household waste, urine, feces and animal carcasses. It was common for garbage and rubbish to accumulate in the streets, sitting for days. Collection and disposal of solid waste was uncoordinated, sometimes privatized so that only the wealthy residents could afford to have their waste collected. In New York, street teams collected garbage, but loaded it on barges which then dumped their loads at sea; in St Louis, Boston, Baltimore and Chicago, refuse was hauled to open dumps near the city's edge.[50] As late as the 1870s, many cities had no public provision for the collection and disposal of household garbage. Ever increasing mounds of waste became a major issue for the industrial city, but the debate was who was ultimately responsible for providing such services – private interests or the city?

Not all land pollution came from the factories. The horse was the principal cause of dirty streets. In 1830 the first horse-drawn buses appeared in New York; at their peak, horses would number 170,000. For the next century, the horse would remain the primary form of transportation for both people and goods, pulling wagons, trolleys and horse cars. Most of the horses were commercial or work animals. They hauled freight and made deliveries. The environmental impacts of the horse were significant. A single horse discharged more than several gallons of urine and nearly twenty pounds of manure each day. The evidence of the horse was everywhere – piles of manure littered the streets attracting swarms of flies and generating stench, and the occasional discarded carcass of a horse that dropped dead while performing its job (see Figure 3.2).[51] Streets could turn into cesspools when it rained; in Paris ladies and gentlemen were assisted in their navigation through streets littered with horse droppings by "crossing-sweepers".[52] The historian Joel Tarr has called the horse the predecessor to the auto, noting that the horse generated many of the same problems attributed to the car: air contaminants, noxious odors and noise. Even more disturbing was the fact that while the horse created environmental problems for the city, urban conditions made life incredibly difficult for the horse. The average streetcar horse had a life expectancy of two years; many horses were whipped and abused by their drivers spurring them on to hauling heavy loads. Many horses died in the open street.

Air pollution

Here was the very heart of industrial America, the center of its most lucrative and characteristic activity, the boast and pride of the richest and grandest nation ever seen on earth—and here was a scene so dreadfully hideous,

Figure 3.2 *Dead Horse in the Streets of New York City, circa 1910*
Source: Library of Congress Prints and Photographs Division, Washington, DC
20540, call number LC-USZ62-54944

> so intolerably bleak and forlorn that it reduced the whole aspiration of man to
> a macabre and depressing joke.
>
> H.L. Mencken writing on Pittsburgh in an essay in the *Boston Herald*

The opening paragraphs of Charles Dickens's novel *Bleak House* (1853) speak
of "smoke lowering down from the chimney-pots, making a soft black drizzle,
with flakes of soot in it as big as full-grown snowflakes – gone into mourning,
one might imagine, for the death of the sun". By the mid 1800s more than a million Londoners were burning softcoal, and winter "fogs" became more than a
nuisance. In 1873 a coal-smoke saturated fog hovered over the city for several
days, causing 268 deaths from bronchitis. By 1905 the term "smog" had been
coined to describe London's combination of natural fog and coal smoke.
By then, the phenomenon was part of London history and the dirty, smoke-filled
"pea-soupers" were legendary.

After the 1830s, many industrial cities relied primarily on using coal as a major
source of energy. Two types of coal are readily found. The first is bituminous or

soft coal, which is high in sulfur; the second type is anthracite coal which is harder and burns more cleanly. In early decades, bituminous was more plentiful and used more frequently, but when it is consumed, much of its residue goes directly into the air. The severity of smoke and particulate matter that is released when coal is burned left its mark on buildings, on laundry and in the lungs of city dwellers. Cities such as Pittsburgh and St Louis relied primarily on bituminous coal. Air pollutants in the industrial city included soot from smokestacks and locomotives; other factories belched chemicals including chlorine, ammonia, carbon monoxide, carbon dioxide, hydrosulfuric acid and methane. Although there were concerns about air pollution, it tended to be regarded as a nuisance and was lower on the environmental agenda for two important reasons. First, the impacts of smoke were the dirtying of buildings and high cleaning prices for clothing, but scientists at the time were unable to link smoke with health problems. In contrast, water pollution had a more immediate and direct link to public health through the spread of infectious diseases, and the rise of bacterial science had showed cause and effect scientifically.[53] Second, smoke was equated with progress, growth and jobs, and the smoky skies were a constant reminder of economic growth and prosperity.

There were some attempts to control smoke. Some US cities banned the use of bituminous coal-burning locomotives from their streets. In 1869 Pittsburgh attempted to ban the construction of beehive coke ovens within the city limits. While some smoke-control efforts had limited success in reducing smoke, they basically failed to make substantial inroads into the problem.[54]

Reforming the industrial city

> The time has arrived when manure heaps, slaughter houses, fat and bone boiling establishments, glue manufactures, outdoor or unsewered privies and all kindred occupations and nuisances cannot be much longer tolerated within the built-up portions of New York or Brooklyn.
>
> *New York City Metropolitan Board of Health,* March 1866.

In the industrial age, economic growth meant progress. In the early years, few people attached importance to preserving the quality of the environment. But there were consequences to unlimited and unregulated growth. Short-term consequences of environmental pollution and the industrial city were numerous: malformation of the bones due to lack of sunshine and a poor diet; skin disease from dirty water; small pox, typhoid and scarlet fever spread through dirt and human excrement in land and water; tuberculosis from bad diet and overcrowded conditions in tenements. Long-term consequences were little understood at the time, but we now know that these include health problems associated with long-term exposure to pollutants and chemicals. City authorities and citizens appeared to tolerate a great deal of filth in their cities and to ignore questions of environmental quality. Many had accepted environmental degradation as the price for economic growth.

Cities became saturated with air, water and refuse pollution, but a growing environmental consciousness and increased concern for sanitation galvanized various reform movements. Pollution became no longer simply a nuisance – it was an unwanted and sometimes dangerous byproduct of industrialization. Major reforms during this time would dramatically change environmental quality (see Table 3.2).

Public health reforms developed out of the dangers of the industrial city. The cholera epidemics of the 1830s and 1840s had social and political ramifications beyond public health, raising fundamental questions about medical treatment, social policy and the livability of the urban environment. In Europe, the 1832 cholera epidemic killed 20,000 Parisians and many in London, helping to convince an enlightened few of the need for a systematic public health system.[55] In 1842 Edwin Chadwick, a commissioner on the Royal Commission to consider public provision for the poor, published his *Report on the Sanitary Condition of the Labouring Population of Great Britain*. It had been preceded by various alarming medical reports on Manchester and London and followed a major outbreak of cholera in 1931–32.[56] Chadwick advocated bringing health and sanitation under a central control, but this threatened the power of many vested, private interests. Ultimately Chadwick's work inspired

Table 3.2 *Industrial reforms and the creation of public utilities*

Pure water supply developed
Street cleaning and garbage pick up
Improved water closet
Stationary bathtubs with water pipes
Water system with running water for every house and apartment
Collective sewage system
Establishment of Board of Health
Establishment of Sanitation Department

new sanitary laws in England and even inspired American health officials to see refuse as a health issue.

What guided thinking before the 1880s was what the environmental historian Martin Melosi has called the "Law of Purification". "Clean" meant whatever the observer could touch, taste, smell or see with the naked eye. If water was clear, odorless and tasteless, it was pure. And moving water was thought to quickly dilute any pollutants, resulting in "harmless" discharges. This "what you see is what you get" approach to pollution analysis lasted until the late 1800s. If the water looked clean it was. If it was cloudy it was impure. That invisible organisms could live and even thrive in a watery environment was beyond imagination. Although the microscope was invented in 1674, it took another two hundred years for scientists to isolate and identify microbes and their link to particular diseases. Finally in the 1890s chemists and sanitary engineers identified sewage-polluted water as the carrier for infectious disease.

The Law of Purification worked when populations were low and the discharge of pollutants were predominately organic rather than inorganic. A large body of water can naturally purify small quantities of sewage and waste, but a large urban population produces so much sewage that bacteria can no longer cope with the heavy load, thus depleting the oxygen levels that fish and plants need to survive. Eventually a river, lake or stream may become eutrophic.

In the US, cities where disease was common began to lose business to those which were healthier. New Orleans, which suffered ongoing outbreaks of cholera, lost out to Chicago; in Memphis, outbreaks of typhoid and dysentery almost destroyed its economy.[57] With more than just civic pride at stake, cities began to consider how to reform sanitation and deal with pollution. The historian M. Christine Boyer commented

> The environmental chaos of the American city became linked in the minds of the improvers to the social pathologies of urban life. Long before poverty, poor housing and slums were thought of as economic and political symptoms, improvers saw a link between environmental conditions and the social order, between physical and moral contagion.[58]

As cities outgrew their local water supplies, they were compelled to develop more extensive public water supplies. Many municipalities attempted to solve the problem through private companies, but there were uneven results. Private companies were often reluctant to provide water for civic purposes like flushing streets clean and the working class and poor could not afford their high prices and so had no access to water. New York City established an ambitious water works program in the 1830s, as they constructed a dam on the Croton River, some 40 miles north of the city, and built aqueducts to convey the water on its long journey.[59] By the 1860s the capacity of the system was 72 million gallons per day, but by the 1880s even this was not sufficient, in part because population growth was so rapid. In 1885, New York City began construction on the New Croton Aqueduct, which would provide 300 million gallons per day. Further urban growth would put strains on even this capacity and so New York decided to access water in the Catskill Mountains, 100 miles to the north. Other cities began to discuss the development of urban water supplies. In addition to the building of new water supply systems, new technologies such as bathtubs, shower-baths and water closets were becoming standard items in middle-class homes. This placed new demands on municipal water supplies. By the late 1880s many US cities developed municipally owned supplies, bringing in water from nearby rivers, lakes and groundwater sources. Cities, once reluctant to spend money on public services, now realized that providing a reliable source of water was both a social and economic imperative. For example, the Citizen's Association of Chicago supported water works projects on the basis that it would improve Chicago's business climate and thus insure the prosperity of their members.[60]

In cities around the Western world, it became a matter of prestige and pride to construct large-scale water and sanitation infrastructure projects. In London, for example, eliminating the 'great stink" of the Thames became a national priority. In Paris, sanitizing the city was one of Baron Haussmann's main concerns when he was Prefect of the Seine from 1853–70. In Athens, bringing adequate fresh water to the city was an urgent political and social issue. However such large-scale engineering projects required major capital investment, an adequate supply of labor, social consensus and political commitment.[61]

In addition to new developments in water supply, many advances took place with regard to water quality (see Table 3.3). In 1882 Robert Koch was able to demonstrate with a microscope that tiny micro-organisms (*tubercle bacillus*) were responsible for the transmission of tuberculosis. In 1884 he identified the cholera bacillus, *Vibrio cholerae*. His work found the bacteria causing bubonic plague, leprosy and malaria and helped pioneer techniques that would allow others to find the germs causing cholera, typhoid fever and anthrax. Finally, the link between organic pollutants, primarily fecal contamination, was understood, and new technologies in filtration emerged to clean the water of such pathogens. Bacterial researchers such as Koch and Louis Pasteur not only established germ theory, but helped to clarify the etiology of various water-borne diseases such as typhoid fever.

Table 3.3 *Outline of water treatment milestones*

312 BCE	First Roman aqueduct
8th century	Distillation
1582	First pump (London)
1652	First American supply work (Boston)
1685	Slow sand filtration
1761	First steam pumping (London)
1829	First large municipal sand filter supply (London)
1849	New York City begins building sewers (125 miles of lines built 1849–1873)
1855	London sewers modernized after cholera outbreak
1858	Year of the "Great Stink" of the River Thames (London)
1885	Water softening
1885	Water bacteriology
1891	Aeration
1895	Iron and manganese removal
1907	Chlorine disinfection

Source: Adapted from Weinstein, M. (1980) *Health in the City*, New York: Pergamon Press, p. 28.

By 1900 more than 21 germs that caused disease had been identified in just 20 years.

In Europe and North America scientists, chemists and biologists pioneered sewage treatment processes. Slow sand or mechanical filters removed many of the pathogens. By the early 1900s water treatment plants were using chlorination to disinfect the water supply. Both the new waterworks and sewage facilities resulted in striking improvement in the city's public health. By the late 1890s, cities in Europe and North America saw dwindling outbreaks of cholera and dysentery. Commitment to full sewage treatment lagged, however, in part because engineers and politicians believed that the risks involved in using streams for sewage disposal were not sufficient to justify the costs of constructing sewage-treatment plants and continued to argue for the natural dilution power of waterways.[62]

An important point is that the focus of most of the reforms from the 1890s to 1930s to drinking water and sewage systems focused primarily on the connection between organic pollution, human wastes and human health, resulting in a concentration on sanitary wastes. The historian Joel Tarr notes that the shift from filth theory to germ theory caused public health authorities to reduce their interest in industrial wastes because they did not normally contain disease germs.[63]

Although there were organic industrial wastes (such as wastes from dairies, canneries and meat packing) industrial wastes also included metals (such as lead, zinc and arsenic), and inorganic pollutants such as dyes, phenol wastes and cyanides. However, at that time, metals and inorganic wastes appeared to have no impact on human health. By placing their highest priority on the potential dangers to water supplies from sewage, public health professionals drew attention away from a concern with industrial wastes. This meant little, if anything, would be altered until the middle of the twentieth century.

Ironically the reforms dealing with human waste shifted the sink for the wastes from the land to the water. Some cities, particularly cities downstream from other cities, saw an increase in infectious diseases such as diarrhea and typhoid.[64] Thus for a while, new reforms in water and sewage sanitation actually increased organic pollutants which in turn increased the risk of pathogens. By the early twentieth century, researchers in sanitary engineering had solved the problem of organic sewage-polluted water with the development of new technologies such as water filtration and chlorination.

In addition to undertaking water reform projects, cities also turned their attention to the refuse problem. Many cities had rudimentary regulations for the disposal of waste. Pigs were used to scavenge; teams of "rakers" were often employed to remove garbage from the city. One of the first problems to be addressed was that of street cleaning. The environmental historian Martin Melosi argues that street cleaning was an easier problem to address than household garbage, because streets were clearly in the "public" domain, whereas households were a question of individual responsibility.[65] In New York City, the Edict of 1866 evicted 299 piggeries, mandated the cleaning up of some 4,000 refuse-heaped backyards, introduced the first watertight garbage cans, established regular rounds for garbage cleanup, and noted that "the well-known capacity of the city to drown in sewage is more than matched by its talent for smothering under a blanket of garbage and refuse".

In the US, in 1887 the American Public Health Association appointed a committee on Garbage Disposal to study the refuse problems in US cities. In addition, civic awareness also played a critical role. In many cities, local civic organizations created pressure for reforms, such as the establishment of departments of public health and other sanitary services, and new laws. The Ladies' Health Protective Association of New York City became a leading force in the fight to bring about sanitation reform in US cities, undertaking projects that included refuse reform, street cleaning improvements and school sanitation.[66]

Perhaps one of the more influential reformers was Colonel George E. Waring, Jr, an engineer and former Civil War officer. He was appointed to the first street cleaning commission of New York City in 1895. Waring created an efficient street cleaning operation. He proposed that household refuse be separated into different receptacles. He also helped to build the first municipal rubbish sorting plant in the US, where salvageable materials were picked out and resold. Waring increased the pay of street cleaners, improved their working conditions and issued them with white uniforms. The sweepers were called "White Wings" (see Figure 3.3). His program for New York City indicates an important shift from viewing the refuse problem only as a question of health to a wider more multifaceted urban problem. Sanitary engineering had now become a formal branch of engineering, and the growth and complexities of urban problems such as sewage, water supply, street cleaning and garbage collection required special training and education.

Reforms to air pollution lagged far behind those of water and land reforms. In part this was because many key policymakers considered economic growth essential, and the toleration of polluted air was so entrenched. For many, a smoky

Figure 3.3 The White Wings, circa 1890s
Source: Library of Congress Prints and Photographs Division, Washington, DC 20540, call number LC-DIG-ggbain-09893

sky was a sign of progress and prosperity. The urban historian Lewis Mumford once said of the mindset of the industrial city "Smoke makes prosperity, even if you choke on it." Opposition to new policy for clean air centered around the argument that smoke meant wealth and jobs. By the turn of the century, however, US cities were reporting "Londoners" – the combination of smoke and fog that had long plagued London. Smoke appeared to be responsible for physical problems such as the increases in pulmonary disease.

In the United States, the Progressive Era brought many local attempts to regulate smoke and air pollution. In particular, local civic leagues and women's organizations voiced their concerns and pressured for policy reform. Many women's clubs were composed of upper-middle-class socially prominent women who had the leisure time to devote to the smoke abatement reform movement. The Ladies' Health Association of Pittsburgh galvanized support for new smoke ordinances in 1892; in other cities like St Louis, Cincinnati, Chicago and Baltimore similar clubs were established. These organizations were often supported by engineers who faced mechanical problems caused by the residual effect of smoke and advocated technical advances to abate smoke. Civic groups such as Chambers of Commerce also took up the crusade as part of "civic pride". Ironically, in 1898, steel magnate Andrew Carnegie delivered a speech to the Pittsburgh Chamber of Commerce urging it to take up the issue of smoke control.[67]

By 1912, many cities had smoke-control ordinances, but many of the ordinances were lax and stiff fines were uncommon as reformers resisted penalizing industry with costly burdens. The equation of smoke with progress caused many political leaders to move cautiously. One of the more far-reaching reforms occurred in Pittsburgh. In 1941, the Pittsburgh Smoke Control Ordinance set a policy goal to eliminate dense smoke as well as other components of air pollution, such as fly ash. This applied to both industrial and domestic sources. Consumers would have either to burn smokeless fuel or to use smokeless technology if they continued to use bituminous coal.[68] Implementation proved difficult but Pittsburgh's air improved considerably by the late 1940s and the city benefited from improved air quality, more sunshine and improved health, as well as savings on cleaning costs and laundry bills.

This cursory discussion of pollution and reforms in the industrial city allows general conclusions to be drawn. First, solutions to one form of pollution often generate new pollution problems in different localities or in different media. Second, both cultural values and scientific knowledge influence pollution policy. Third, nineteenth- and early twentieth-century pollution reforms relied upon a technological solution or fix. There was little effort to question or prohibit the production of pollution, rather efforts and technology focused on how to clean it up once it was in the environment.

The urban public parks movement

The deteriorating industrial city prompted numerous reformers, including landscape architects and planners to revision the city and nature–society relationships. The urban public parks movement, which emerges in both Europe and the US during the mid to late nineteenth century, is a reaction to the problems of the industrial city as planning "visionaries" offer designs for a reconstructed relationship to nature through urban space. Although most cities had open land within city limits (vegetable gardens, squares, commons) these were not often formalized as public space or recreational areas.[69] In both Europe and in America, the ideology of the public park began with a recognition of the importance of open spaces to the health and vitality of the urban population.[70] A constellation of factors provided the genesis of the urban public parks movement in the US: worsening urban conditions and the link with public health, a concern for intellectual and moral improvement, and a challenge to the nation's self-esteem. Urban parks would be aligned with democratic ideals and would be accessible for all socioeconomic classes: parks were to be for the public.[71]

One of the most important of these visionaries is Fredrick Law Olmsted, designer of Central Park in New York and many other parks and park systems, residential developments and college campuses in numerous North American cities. His legacy continues to impact many cities and his planning ideology informed a generation of park planners and landscape architects.

Fredrick Law Olmsted was born on 26 April, 1822 in Hartford, Connecticut. He was the son of a dry goods merchant. As a young man, he studied civil engineering but struggled until his mid-forties to define his professional identity, shifting from farming, to journalism and then to landscaping.[72] In 1850, when he was 28, he went to England and wrote a book about farming methods, *Walks and Talks of an American Farmer in England*. In 1852, he went to work as a reporter for the *Times*, and was assigned to write about the South. His experience there sparked his interest in learning more about the democratic condition of society. He moved from writer to landscape architect when he was hired as the New York City park superintendent before anybody had any idea what the Park might look like.[73] New York City had set aside some 750 acres of land on the outskirts of town as the site of a new city park. The site itself was a mixture of swampy areas, rocky outcroppings, and contained numerous heaps of cinders, bricks and other rubbish. The park would be made, not found. As one report noted: "never was a more desolate piece of land chosen for a pleasure ground".[74] In 1858, Olmsted and fellow landscape architect Calvert Vaux, a trained architect, worked together for several months and won the prize for the design for Central Park.

The design for Central Park reveals a very different perspective on nature–society relationships than Versailles or the Boboli Gardens. While Olmsted was influenced by European gardens, he also sought an American derived model that sprang mainly from an anti-urban ideal. Several design elements are worth mentioning. First, Olmsted and Vaux moved away from the formalized, rationalized geometric design of European gardens and beds and instead stressed curvilinear paths with more native plantings. There would be no manicured, clipped animal topiaries in an Olmsted design. Parks were not to provide pictures or allegories; plantings should be permanent and local. Olmsted resisted regimentation and a central *allée*, as characterized Versailles or the Luxembourg Gardens. Central Park belonged not to a palace, but to the people. The plan is perhaps best characterized as seemingly disordered and unstructured. Lakes and ponds are nestled in their own places and are not part of a grand canal or waterway. Main spaces are not set apart, but merge with other areas. The design reflected a more organic view of the world and sought to offer opportunities for urban dwellers to "reconnect" with nature, to find spirituality in the green trees, grass and vegetation that contrasted with the bleak industrial city. Curvilinear pathways contrast with the "wearisome rectangularity of cities".[75] It was intended to be an adaptable place. The park was to be the place of numerous activities – roller-skating, ice-skating, rowing, puppet-shows, walking, and carriage driving which had both separate and intersecting points. Indeed, almost as soon as the park opened in 1858 the skating pond produced a skating craze.

Olmsted believed parks were the "lungs of the city" and criticized the industrial city for its lack of sunlight, the absence of trees and open spaces. Urban residents, he argued, needed an opportunity to "supply the lungs with air screened and purified by trees, and recently acted upon by sunlight, together with the opportunity and inducement to escape from conditions requiring vigilance, wariness and activity toward other men".[76] Parks were therefore not only aesthetically important, they were linked to public health. Olmsted continues

> We want a ground to which people may easily go after their day's work is done, and where they may stroll for an hour, seeing, hearing, and feeling nothing of the bustle and jar of the streets, where they shall, in effect, find the

city put far away from them. We want the greatest possible contrast with the restraining and confining conditions of the town which will be consistent with convenience and the preservation of good order and neatness."[77]

Olmsted's perspective on nature–society relationships is also connected to wider socialist ideals and the intellectual movement or Romanticism of the time. Romantic idealism or the Transcendentalist movement was emerging as an East Coast intellectual movement led by writers such as Henry David Thoreau and poets such as Ralph Waldo Emerson and William Cullen Bryant. Many had traveled to Europe, visiting parks and gardens. Although most of these parks were originally restricted to use by royalty, by 1850 many were open for the enjoyment of all the people. This was in contrast to many US cities, where only a few cities – Philadelphia, Washington, DC, Savannah, Boston and New York – had set aside public spaces. A number of American writers became proponents for public parks. Romantic idealism looked to the natural world to inspire and to fill the spiritual void present in the dirty, gritty industrial, mechanical city. Romantic idealism stressed a return to "harmony" with nature and a new-found appreciation for it. Transcendentalists attributed virtues to trees, meadows and lakes and believed that these virtues could be duplicated by human ingenuity and design, thus influencing park design theory. In a sense, Central Park "transcended the city". Ironically, this intellectual perspective emerged in an industrial urban context.

Olmsted's design embraced Romantic ideals – rejecting the grid system in favor of wandering paths. Olmsted and Vaux created informal naturalness. Recognizing that recreating the wilderness would be difficult if not impossible within the urban boundaries of the park, Olmsted opted instead for the pastoral and picturesque. This style called for a composition of smoothness, harmony, serenity and the occasional reminder of the awesome grandeur of a mountain, crevasse, waterfall, or lake. The design was to *suggest* nature in places where nature was not actually provided. He favored irregular planting, which suggested the feeling and idea of distance, and tried to avoid the use of flowers, which he felt revealed the hand of man.[78] And he rejected straight lines, hard edges and right angles – all of which were associated with machines and regimentation. Instead, he opted for softer, rounder, more fluid geometries associated with the contours of landscapes and the organic structure of vegetation: it was a romantic rejection of the rational.

As much as Olmsted was influenced by transcendentalism, he also embraced progressivism. Olmsted was also concerned about creating a more democratic social order and his earlier writings had been influenced, to some degree, by communitarian thinking. His park design was connected to his political views. In failing to provide opportunities for public recreation, American society had set up barriers of class and caste. Thus an urban public park would not only foster social freedom, it would also act as an agent of moral improvement by providing organized activities and sports.[79] In park design, this translated into building a better park as a way of helping to shelter commonplace civilization. Parks were more than scenery: they were social spaces. He believed that the immigrant factory worker or unattended child, with no access to parks or recreation, or too much leisure time at the weekends, could become so disillusioned as to threaten the political system or the social order. Hence the parks provided an "outlet" for all social classes. His designs attempted to provide space for all classes and he viewed recreational opportunities in the parks as providing outlets that would "counteract the evils of town life". Thus in the Olmstedian ideology, parks served as democratic playgrounds. And they were. The parks built during the nineteenth century were, by and large, used mostly by children and the urban working classes.

Olmsted's legacy is widespread around US cities. He designed Prospect Park in Brooklyn, and Boston's Emerald Necklace (linking various small parks together along a connecting trail), created a model suburban village at Riverside Illinois, planned Mount Royal Park in Montreal and Bell Isle Park in Detroit, created the campus plan for Stanford University in Palo Alto, California, designed the space around the US Capitol and laid out George W. Vanderbilt's mountain estate, The Biltmore, near Asheville, North Carolina. He also contributed to preservation plans for both Yosemite and Niagara Falls. His legacy is found not only in the actual parks and other places he built, but in an ideology that helped to generate the broader urban park movement during the Progressive Era in America. Often spearheaded by local elites and civic organizations, the movement would bequeath to US cities such parks as San Francisco's Golden Gate Park, Chicago's South Park system, and parks in Philadelphia, Baltimore, Boston, Buffalo and New Orleans. While these parks have evolved over time to accommodate new demands and needs, the urban parks movement set aside vast areas for parks and recreational spaces that would likely have succumbed to the relentless march of the gridiron. The park movement, born in the city, eventually expanded to preserve threatened scenery and helped to establish a constituency for preservation, eventually providing both intellectual and popular support for the creation of state park systems and the national park system.

The emergence and development of an urban parks movement represents an important shift in how society values the natural world within the urban context.

It was a powerful force in the evolving redefinition of American urban form and culture and it laid the conceptual foundation for twentieth-century park development and evolution. And yet it is also important to realize that all urban parks (of any size and function) are cultural landscapes: they are "managed" nature that has been engineered, planted and continually recreated. Under Olmsted's supervision, an army of laborers blasted tons of rock, moved thousands of cartloads of earth, refashioned swamps into lakes and ponds, planted turf, trees and shrubs. As the environmental historian David Schulyer notes, "it is thus a subtle irony ... that, like the city that surrounds it, Central Park is totally a man-made environment".[80]

Garden Cities

In 1898, Ebenezer Howard published his book *Garden Cities of To-morrow*. According to Howard, cities in Europe and America had become too densely populated and suffered from "foul air, murky skies, slums and gin palaces". He was fearful of the consequences for society if old cities – and the social conflicts and miseries they embodied – continued. His vision did not seek the amelioration of the old industrial city, but a wholly transformed urban environment.[81] His vision of Garden Cities sought to create new, self-contained cities of some 500,000 people on some 12,000 acres, which combined the best features of urban life (social opportunity, places of amusement) with that of the "country" (land, fresh air and an abundance of water and sunshine). This was an ideal model, a blueprint for a new type of city. He would reorient the spatial balance between the built form and green space/open space, thus reforming the physical environment and revolutionizing society. This would "restore the people to the land" as he believed that "human society and the beauty of nature are meant to be enjoyed together".[82] His vision contained two important design elements that would have a lasting legacy on urban planning: zoning and greenbelts.

Howard's map of a Garden City is highlighted by the division of land into different uses (Figure 3.4). This is one of the earliest articulations of "zoning". His plan locates all residential areas away from the belching smokestacks of the factory. In addition, he has incorporated space for urban farms, to grow food to supply the city. From the center node out to the radiating smaller nodes, he has planned for railway transportation to link all nodes together. His plan also called for "buffer zones" between each of the distinct land uses. This, too, is one of the earliest articulations of a "green belt" and has become widely adopted in many metropolitan areas in the US and Europe.

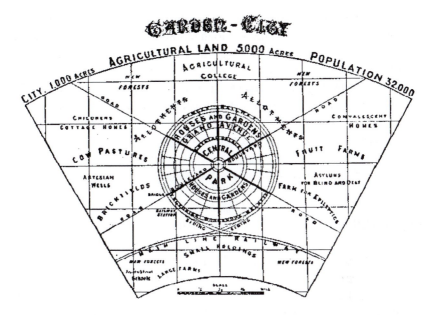

Figure 3.4 Howard's Garden City. This diagram appeared in Garden Cities of To-morrow, originally published in 1898

In England, Howard and his supporters founded two cities, Letchworth (1903) and Welwyn (1920). In the US more than a dozen "Greenbelt Cities" were undertaken, including Greenbelt, Maryland; and in Australia several were established. However Howard's actual Garden Cities failed to inspire the urban revolution he hoped.

Howard looked at cities in a regional, rather than an isolated framework and his vision of a Garden City model represents a major step forward in terms of organizing urban space. His vision spoke of a city where people were reunited with nature. Geographer and planner Robert Freestone has observed that it is possible that Howard's physical innovations of Garden Cities had little impact on the way in which people actually lived, nevertheless his ideas and theories influenced a generation of planners throughout the twentieth century.[83] New visions for reconnecting nature and society in the urban fabric would also appear with the emergence of the modern environmental movement of the late 1960s and 1970s.

Conclusions

The urban parks movement and the intellectual legacy of Ebenezer Howard's Garden Cities provided a counterbalance to the pervasive environmental degradation of the industrial city. New forms of urban design can offer a positive revisioning of the city–nature dialectic.

Guide to Further Reading

Beinart, W. (2005) *Environment and History*, London: Routledge.

Brimblecome, P. (1987) *The Big Smoke: a History of Air Pollution in London since Medieval Times*, London: Methuen.

Cronon, W. (1992) *Nature's Metropolis*, New York: W.W. Norton.

Hall, P. (2002, 3rd edn) *Cities of Tomorrow: An Intellectual History of Urban Planning and Design in the Twentieth Century*, Oxford: Blackwell.

Meller, H. (2001) *European Cities, 1890–1930s: History, Culture and the Built Environment*, Chichester, New York: Wiley.

Melosi, M. (1981) *Garbage in the Cities: Refuse, Reform and the Environment, 1880–1980*, College Station: Texas A&M University Press.

Melosi, M (2000) *The Sanitary City: Urban Infrastructure in America from Colonial Times to the Present*, Baltimore: The Johns Hopkins University Press.

Mosley, S. (2001) *The Chimney of the World: A History of Smoke Pollution in Victorian and Edwardian Manchester*, Isle of Harris: White Horse Press.

Tarr, J. (1996) *The Search for the Ultimate Sink: Urban Pollution in Historical Perspective*, Akron, OH: University of Akron Press.

Tarr, J. (2003) *Devastation and Renewal: An Environmental History of Pittsburgh and Its Region,* Pittsburgh: University of Pittsburgh Press.

4 Contemporary urbanization and environmental dynamics

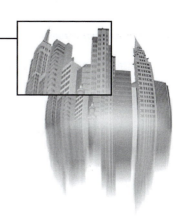

We are in the midst of the Third Urban Revolution. The first began over 6,000 years ago with the first cities in Mesopotamia. These new cities were less the result of an agricultural surplus and more the reflections of concentrated social power that organized sophisticated irrigation schemes and vast building projects. The first urban revolution, independently experienced in Africa, Asia and the Americas, was a new way of living in the world. The Second Urban Revolution began in the eighteenth century with the linkage between urbanization and industrialization that inaugurated the creation of the industrial city and unleashed unparalleled rates of urban growth and environmental transformation. We are currently in the Third Urban Revolution, a complex phenomenon that began in the middle of the twentieth century and is marked by a massive increase, in both absolute and relative terms, in urban populations, the development of megacities and the growth of giant metropolitan regions, global redistribution of economic activities as former manufacturing cities decline and new industrial cities emerge elsewhere. Services, especially advanced producer services, have become the cutting-edge of rapid urban economic development. Consequently, the global urban network is in transition. Urban landscapes are revalorized and devalorized at an often bewildering pace: central cities have characteristically become sites of new urban spectacle; inner cities are pockmarked by sites of gentrified renaissance as well as rampant poverty and criminality; inner suburbs show the first inklings of decline, exurban development continues apace as gated communities and mixed-use developments sprawl into the former countryside. We are in the throes of a revolution that we are only just beginning to see, name and theorize. The new lexicon which has emerged to describe cities – "post-modern", "global", "networked", "hybrid", "splintered" – offers some purchase on the rich complexity and deep contradictions of the Third Urban Revolution, but much remains to be said and done before we can make any sense of the new forms of urbanism which characterize the twenty-first century.[84]

In this chapter we will consider some of the major environmental resonances of this dramatic shift. Let us begin by noting three main aspects of global urban change.

Urban dynamics

The first is that throughout the twentieth century, the world has become increasingly urban. In 1900, only 10 percent of the world's population lived in cities. In little more than a century, there has been a major demographic and social shift with important consequences for environmental quality. Tables 4.1, 4.2 and 4.3 document urbanization trends by region and for selected countries. For example, in 1900, 40 percent of the US population lived in urban areas; today nearly 80 percent do. In Australia, more than 90 percent of the population lives in the six urban areas – Adelaide, Brisbane, Canberra, Melbourne, Perth and Sydney. Urbanization is occurring rapidly in the developing world. In 1900 only 10 percent of Mexicans lived in cities; by 2000 75 percent did. In the Sudan, just 6.8 percent of the population was urban in 1950; by 2000 more than a third of the population was urban. Currently three-quarters of global population growth occurs in the urban areas of developing countries.[85]

A constellation of factors set in motion particularly high levels of urbanization in the developing world. In the post war years (1950–80), large-scale development projects, funded in part through the World Bank, were often channeled to create urban infrastructure for an industrial economy. Development projects that were focused in cities included water and sanitation systems, electrical systems, the building of roads, factories and warehouses, port facilities, and the construction of government infrastructure such as courthouses and parliaments. Although there were many rural development projects such as dams and large highways, significant amounts of development loans and funds ended up concentrated in urban areas.

Table 4.1 *Percentage Urban, 1950–2030*

	1950	1975	2000	2010	2030
World	29.00%	37.00%	46.00%	50.80%	59.90%
MDCs	52.10%	66.15%	73.20%	75.20%	80.60%
LDCs	18.10%	26.80%	40.30%	45.50%	56.10%

Source: Data, United Nations World Urbanization Prospects: the 2005 Revision Population Database.
http://esa.un.org.unup

Table 4.2 *Urban Population in billions 1950–2000*

	1950	1975	2000
World	0.75	1.54	2.86
More developed countries	0.45	0.73	0.92
Less developed countries	0.31	0.81	1.96

Source: Data, United Nations World Urbanization Prospects: the 2005 Revision Population Database.
http://esa.un.org.unup

The high levels of urban primacy in many developing countries also reinforced growth in selected urban centers. The net result was the encouragement of massive rural to urban migration. To those living in rural areas, cities appeared places of economic opportunity, and since many rural areas remained without water or electricity or sanitation, cities also seemed to be better places to live. In many developing cities, the unprecedented population growth often exceeded the formal "coping capacity" of the city. In effect, the city was overwhelmed with too many migrants and not enough housing, schools, or health services. During the 1970s and 1980s, for example, it was estimated that Mexico City grew by some 300,000 people each year. Any city would find it difficult to supply housing

Table 4.3 *Percentage population that is urban for selected countries, 1950–2010*

	1950	1975	2000	2010
Brazil	36.2	61.7	81.2	86.5
Ethiopia	4.6	9.5	14.9	17.4
India	17	21	27.7	30
Iran	27.5	45.7	64.2	69.5
Kenya	5.6	12.9	19.7	22.2
Mexico	42.7	62.8	74.7	77.3
Peru	41	61.5	71.6	75
Philippines	27.1	35.6	58.5	66
Sudan	6.8	18.9	36.1	45.2
Thailand	16.5	23.8	31.1	34

Source: Data, United Nations World Urbanization Prospects: the 2005 Revision Population Database.
http://esa.un.org.unup

Table 4.4 *Urbanization for selected cities, 1950–2010*

City population in millions	1950	1975	2000	2010
Tehran	1	4.2	6.9	7.8
Addis Ababa	0.392	0.92	2.5	3.4
Delhi	1.3	4.4	12.4	16.9
Lima	0.97	3.6	6.8	8.1
Bangkok	1.3	3.1	6.3	6.9
São Paulo	2.3	9.6	17	19.5
Rio de Janiero	2.9	7.5	10.8	12.1
Khartoum	0.18	0.88	3.9	5.1
Manila	1.5	4.9	9.9	11.7
Nairobi	0.13	0.67	2.2	3.3
Mexico City				

Source: Data, United Nations World Urbanization Prospects: the 2005 Revision Population Database.
http://esa.un.org.unup

and health services to an additional 300,000 people yearly. As a result, many developing cities grew in informal ways, embodied in the creation of a unique form of urban space: the shanty town.

The second trend is that, throughout the world, cities have continued to grow larger. In 1800 there were only two cities – London and Beijing – that had more than 1 million inhabitants. By 1900, there were thirteen. Today, there are several hundred that exceed 1 million in population and there are more than 35 that have more than 5 million inhabitants. By 2015, it is estimated that there will be 360–400 cities with at least 1 million inhabitants. The average size of cities has grown dramatically. Table 4.4 shows the rapid urbanization rates for selected cities in the developing world.

One of the major shifts has been in the geography of the world's largest cities. Table 4.5 shows the ten most populous cities in 1950, 2000 and projected for 2010. Note that most of the cities listed in 1950 are located in the industrialized, highly developed regions of Europe and North America. In 2000, many of those cities were no longer among the largest; they have been supplanted by massive cities in the developing world such as Mexico City, São Paulo, Calcutta and Mumbai.

Table 4.5 World's 10 most populous cities 1950–2010 (in millions)

Population 1950		Population 1975		Population 2000		Population 2010	
New York	12.3	Tokyo	26.6	Tokyo	34.4	Tokyo	35.8
Tokyo	11.2	New York	15.8	Mexico City	18.0	Mujmbai	20.4
London	8.3	Shanghai	11.4	New York	17.8	Mexico City	19.8
Paris	5.4	Mexico City	10.6	São Paulo	17.0	São Paulo	19.2
Moscow	5.3	Osaka Kobe	9.8	Mumbai	16.0	New York	19.1
Shanghai	5.3	São Paulo	9.6	Calcutta	13.0	Delhi	18.2
Rehin Ruhr	5.2	Buenos Aires	9.1	Shanghai	12.8	Jakarta	15.4
Buenos Aires	5.0	Los Angeles	8.9	Buenos Aires	12.5	Calcutta	15.4
Chicago	4.9	Paris	8.6	Delhi	12.4	Dhaka	15.1
Calcutta	4.4	Beijing	8.5	Los Angeles	11.8	Lagos	14.0

Source: United Nations Department of Economic and Social Affairs/Population Division, *World Urbanization Prospects: the 2003 Revision.*

Urban regions

There have been a number of attempts to model land use around a city.

A model of intensity around the city was first developed as far back as 1826 by a German landowner Johan Heinrich von Thunen (1783–1850). Von Thunen noted that there was a pattern to land use around a city: in particular more intensive uses were closer to the city. Von Thunen postulated a general model of land use around a city situated in a flat plain with homogeneous fertility and transportations costs. Farmers' costs were based on land costs and transport costs. Since farmers paid less transport costs closer to the city, land costs tended to be higher. Only farmers growing the more intensive crops, with high returns, could afford the land closer to the city. The net result was a concentric ring pattern with more intensive agriculture closer to the city.

Von Thunen's model was developed at a local scale with reference to agricultural land use. The historian William Cronon developed his model at a larger spatial scale in relation to the commodification of nature. In his 1991 book *Nature's Metropolis* he examines the relationship between Chicago and its hinterland from 1850 to 1890. He shows how the physical world was turned into a commodified human landscape as grain, lumber and meat production transformed

Continued

prairies and woodlands into the physical basis for the city's growth and development. Merchants, railway owners and primary producers transformed the "wilderness" into a humanized landscape that was the basis for the city's impressive economic growth. Cronon's work demonstrates that urban economic growth draws heavily on a physical world.

We can also imagine a model pitched at the global level. At its heart is the concentrated urban center of the global cities. Around this urbanized landscape is a suburban semi-periphery and beyond that a periphery of agricultural, forest and other areas. We can model the world as one giant urban region of dense urban centers, a suburban semi-periphery and rural/wilderness periphery with flows and interconnected relations between all three. The insights of the von Thunen and Cronon models can be recast at a global level.

Sources: Cronon, W. (1991) *Nature's Metropolis: Chicago and the Great West*, New York: Norton.
Von Thunen, J. (1966) *Isolated State*, Oxford: Pergamon (an English edition translated by C. M. Wartenberg, edited with an introduction by P. Hall).

The third trend is the creation of giant urban regions and megacities. A number of recent studies suggest that large urban regions are the new building block of both national and global economies.[86] Scholars have identified globalizing city regions in which most of urban and industrial growth is concentrated. Peter Taylor describes the world economy as structured around an archipelago of global city regions.[87] In the developed world these city regions are the loci of control and command functions with a heavy concentration of advanced producer services such as banking, advertising and business services. In the developing world these are the site of multinational corporation investments and new techniques of manufacturing as well as centers of service industries. Three giant urban regions have been identified in Asia Pacific: Bangkok (11 million population), Seoul (20 million) and Jakarta (20 million), which have between 35 percent and 25 percent of all foreign direct investment into their respective countries and constitute between 20 percent and 40 percent of their respective national gross domestic product. In China, for example, the three city regions of Beijing, Shanghai and Hong Kong constitute less than 8 percent of the national population, yet they attract 73 percent of the foreign investment and produce 73 percent of all exports. China is less a national economy than three large urban economies.

In the US there are 10 megapolitan regions, defined as clustered networks of metropolitan regions that either have populations of more than 10 million or will exceed that number, on current growth projections, by 2010. Collectively, these large city regions constitute only 19.8 percent of the nation's land surface yet comprise 67.4 percent of the population, and approximately three-quarters of all predicted growth in population and construction from 2010 to 2040.[88]

One of the more visible aspects of global urbanization has been the rise of megacities, large urban agglomerations with more than 10 million inhabitants. They are a recent addition to the urban scene. In 1950 only New York and Tokyo had populations of more than 10 million. Today there are more than 20 megacities, with the majority of them in Asia and Latin America (see Table 4.6). By 2015 there may be 26 megacities, 22 in developing countries. Lagos is one example. In 1995, Lagos became the world's 29th largest urban agglomeration with 6.5 million inhabitants. In 2000, it was the 23rd largest with 8.8 million. In 2002 Lagos became the first sub-Saharan African megacity when it topped 10 million. With an annual population growth rate of 9 percent, it is one of the fastest growing cities in the world. By 2015 it may have as many as 14 million residents. Consider São Paulo, Brazil. In 1950, it was a city of some 2.3 million inhabitants. By 1975 its population had quadrupled to 9.6 million; it has since doubled: today estimates range between 16–19 million.

Table 4.6 Megacities, 2005 (population in millions)

Asia	South Asia	South East Asia	Latin America	North America	Europe/ Russia	Africa
Tokyo (35.1)	Mumbai (18.2)	Jakarta (13.2)	Mexico City (19.4)	New York City (18.7)	London (12.1)	Lagos (10.8)
Osaka (11.2)	Delhi (15.0)	Manila (10.6)	Rio de Janeiro (11.5)	Los Angeles (12.3)	Moscow (10.6)	Cairo (11.1)
Seoul (10.1)	Kolkata (14.2)		Buenos Aires (12.5)			
Shanghai (14.5)	Dhaka (12.4)		São Paulo (18.3)			
Beijing (10.7)	Karachi (11.6)					

Source: United Nations Department of Economic and Social Affairs, 2005, "Urban agglomerations, 2005". Accessed at www.unpopulation.org March 9, 2007.

In the rest of this chapter we will consider five environmental aspects of this third wave of urban transformation; the impact of giant urban regions and megacities, postindustrial cities and brownfields, urban sprawl, new industrial spaces and shantytowns.

Urban regions and megacities

The largest city region in the US is the urbanized northeast seaboard, a region first named by Jean Gottmann as Megalopolis. Megalopolis stretches from just south of Washington, DC, through Baltimore, Philadelphia and New York to just north of Boston. It is responsible for 20 percent of the nation's Gross Domestic Product. In 1950 Megalopolis had a population of almost 32 million people. One in four of all US residents lived in this region. By 2000 the population had increased to almost 49 million. This small area of just over 52,000 square miles with only 1.4 percent of the national land surface still contains over 17.3 percent of the nation's population. Despite the national redistribution of the US population Megalopolis continues to remain a significant center for the nation's population with almost one in six of the US population living in this single giant urban region. In 1950, the average population density was 610 people per square mile. By 2000 it was 931. The environmental impact of this population increase is enormous: more people driving more cars to more places; more people running dishwashers, flushing toilets and showers; more people in more and ever bigger houses. Megalopolis is arguably one the most environmentally impacted regions in the US subject to the constant, mounting stress of a rising population with an ever-growing list of needs and desires.

Consider automobiles. Applying the standard estimates of one car for every five people in 1950 and one car for every two people in 2000 yields a total of 6.4 million cars in 1950 and 24.25 million cars in 2000. In the same surface area, the number of cars has almost quadrupled. And this total does not include the buses, cars and trucks passing through this region from the outside. There are now, at the very lowest estimate, over 24 million autos releasing exhaust, needing roads and requiring parking spaces. The landscape has been redesigned to give these cars the space and freedom to move throughout the region.

And water usage: in 1950 daily per capita water withdrawals for the US totaled 1,027 gallons, and by 1995 (the latest available statistic) this figure had increased to 1,500 gallons per capita per day. In Megalopolis, not only has the population increased but the daily withdrawal has increased by 50 percent per head. These statistics need to be treated with some care: national statistics tend to overestimate rates of withdrawal since per capita water usage is less in urban areas than rural areas. However, while the absolute amounts are only very rough estimates, their direction is clear. Total water withdrawal increased in the region by over 150 percent from 1950 to 2000.

A similar picture emerges for municipal solid waste generation. In 2000 4.3 lb was generated per person per day. By the end of the twentieth century the population of Megalopolis generated approximately *one hundred thousand tons of garbage per day*. Consider the case of New York City, which each day generates approximately 12,000 tons of municipal waste and an equal amount of waste from businesses that are collected by private companies. Since the closing of the Fresh Kills landfill in Staten Island in 2002, the city now has an elaborate system for the transfer and disposal of the municipal waste. Trash is collected and hauled by 550 trash trucks to transfer stations in the city and New Jersey, and then transported for incineration in Newark and dumping in landfills in Pennsylvania, New Jersey and as far south as Virginia. Trash truck journeys have increased congestion and raised the pollution levels along such routes as Canal Street by as much as 17 percent. In some of the receiving areas taxes from landfill companies pay as much as 40 percent of the school budget. Megalopolis is one giant waste-generating, waste-disposal region.

Whatever the measure, it is the same story of increasing population growth in association with increased affluence and spiraling consumption producing a greater environmental footprint and increased strain on the natural systems that sustain and nurture life. As more population crams into the region, an incredible cost is exacted by the environmental transformation that is wrought. Close to 50 million people, with the greatest environmental impact per head in the history of the world, now live in Megalopolis.

Megacities – defined as cities with populations of at least 10 million – also impose a heavy environmental toll. Consider the case of Mexico City. In 2005, the city's population was an estimated 19.4 million, of which some *10 million* live in shanty settlements around the city. Some 40 percent of Mexico City's population is without sewage. In 25 percent of the children, higher than normal lead levels were reported in their blood due to exposure to air-borne lead from both leaded gasoline and the use of leaded paints. And between 1970 and 2000, more than 70 percent of city trees died, primarily from ozone poisoning.

Megacities are magnets for people, organizations, of and economies, and the fulcrum of many countries' social and economic dynamics. Megacities are not just cities of a large size, they are a new distinctive spatial form of social organization, economic production and political governance.[89]

We can identify both direct and indirect environmental consequences attributed to the sheer magnitude of megacities. For example, megacities are so large, the volume of pollutants is very high, and millions of residents are at risk. Fifteen megacities are located on the coast including Tokyo, Shanghai, Tianjing, Bangkok, Jakarta, Manila, Mumbai, Karachi, Istanbul, New York, Los Angeles, Buenos Aires, Rio de Janeiro and Lagos. Coastal megacities have special concerns such as land-use conflicts in coastal areas as well as problems such as coastal erosion, salt water intrusion and freshwater shortage and the depletion of fishery resources.[90] For example, rapid demographic and economic growth in Jakarta has created water pollution, coastal erosion, mangrove destruction and the intrusion of salt water into the freshwater supply. In Mumbai, high polluting industries including chemicals, fertilizers, iron and steel, and petrochemicals have released semi-treated or untreated waste material into the coastal waters, degrading the beaches and impacting the tourism industry. In Buenos Aires, untreated sewer water has been dumped into the River Plate and ultimately into the seas, creating serious coastal environmental problems for the city.

Air pollution can be particularly severe in megacities (see Table 4.7). Motor vehicle traffic is a significant source of air pollution in all of the megacities; in nearly half of the world's megacities it is the single most important source. Emissions from power generation are also a problem. In China, Beijing and Shanghai are dealing with high levels of sulfur pollution that stem from the use of coal as a major energy source. China is now home to 16 of the world's 20 most air-polluted cities and demand for fossil fuels (particularly coal) has only increased.

On a positive note, some megacities have made vast improvements in air pollution. São Paulo is economically the most important city region in Brazil. Emission restrictions have helped reduce pollution from many industrial sources. The Brazilian government has encouraged alcohol-fueled vehicles and now more than 60 percent of São Paulo's vehicles run on gasohol – a mixture of gasoline and ethanol. This has helped reduce some air pollutants, although alcohol combustion does generate aldehydes.

Megacities consume vast amounts of resources. Bangkok, Manila and Mexico City, for example, consume vast quantities of water and are already dangerously depleting their groundwater supplies. Bangkok has subsided more than 14 centimeters in the last 50 years due to excessive withdrawal from natural aquifers, thus exacerbating urban flooding problems and causing damage to buildings and infrastructure.[91]

Table 4.7 Air pollution in megacities

City	SO_2	PM	Lead	CO	NO_2	Ozone
Mexico City	Severe	Severe	Heavy	Severe	Heavy	Severe
Buenos Aires	Low	Heavy	Moderate	Severe	Severe	Severe
Beijing	Severe	Severe	Moderate	Low	Moderate	Heavy
Cairo	Low	Severe	Severe	Heavy	Low	Low
Seoul	Severe	Severe	Moderate	Moderate	Moderate	Moderate
Jakarta	Moderate	Severe	Heavy	Heavy	Moderate	Heavy
São Paulo	Moderate	Heavy	Moderate	Heavy	Heavy	Severe
Manila	Low	Severe	Moderate	No data	No data	No data
Delhi	Low	Severe	Low	Low	No data	No data
Kolkata	Low	Severe	Low	No data	No data	No data
Shanghai	Moderate	Severe	No data	No data	No data	No data
New York	Low	Low	Low	Moderate	No data	No data
Tokyo	Low	Low	No data	Low	No data	No data

Severe: serious levels of pollution that exceed World Health Organization guidelines by a factor of two or more.
Heavy: WHO guidelines exceeded by up to a factor of two.
Moderate: WHO guidelines are exceeded.
Low: WHO guidelines are normally met.
Source: World Health Organization, various reports online from 1998–2006

Changes in a Bangkok slum, Klong Thoey

The vulnerability of shanty dwellers is often compounded by the fact that they do not have legal entitlement to the land. In the passage below, small changes occur in one Bangkok slum, Klong Thoey.

To qualify for a government school, a child has to have a birth certificate, but to get such a document the parents have to live in a registered house; the problem, of course, was that none of the slum houses were, or could be, registered, since that would imply legal ownership of the land.

The great majority of Klong Thoey's families accepted this as merely another fact of slum life, along with crime, the lack of sanitation and the constant threat of eviction. Prateep's mother was an exception. She was determined to give her daughter at least the rudiments of an education, and to do so she was willing to somehow scrape together the necessary money to send her to a cheap private school on the fringe of the slum. By the time she was 15, she had saved enough

Continued

to enroll in an evening school for adults, where she managed to complete six years of study in only two and a half ... Finally, certificate in hand, she turned one small room of the family shack into a sort of daycare center for young children of working parents, some of whom paid her a baht a day for the service. [...]

The Port Authority informed Prateep and some two thousand of her neighbors that the area in which they lived—officially designated as Block 12—was required for "development" and that they would have to leave ... But even when squatters have inhabited an area for upward of 30 years, as many in Klong Thoey had, they have no legal rights in Thailand. No one paid much attention until a reporter from the *Bangkok Post* happened to hear about it, and saw the possible makings of a good story. [...]

Cast in the role of the villain, the Port Authority compromised. The residents of Block 12 would still have to move, but they could relocate to some empty land not yet needed, further back in the slum. The move turned into a show of unprecedented slum solidarity. The three-hundred-odd families involved selected representatives to divide the new area into housing sites, and when the time came they helped one another take down the old shacks and reassemble them. Half an acre was reserved on which to build a real school for the young teacher, now regarded as a leader.

Source: Warren, W. (2002) *Bangkok*, London: Reaktion Books, pp. 134–136.

In addition to the environmental impact that cities have directly, as polluters, there are also indirect consequences of megacities. As megacities grow, the peripheries of these cities enlarge, consuming agricultural land, forests and wetlands. Dhaka, Bangladesh, is forecast to become the fifth largest city in the world by 2015, with a forecast population increase of 8.5 million in the next 15 years. Yet Dhaka is bounded on the west and south by the flood plain of the Burhi Ganga River and on the east by the flood plain of the Balu River. Both areas are flooded up to four months of the year. Land above the flood plain is high value agricultural land but is rapidly being converted to urban uses as Dhaka expands.[92]

Although megacities have mega-environmental impacts, it is important to note that many smaller cities have environmental degradation problems that are often more severe. According to the Blacksmith Institute, the 10 most polluted places in the world include Chernobyl (Ukraine), Haina (Dominican Republic), Kabwe (Zambia), La Oroya (Peru), Linfen (China), Norlisk (Russia), Dzerzinsk (Russia) and Ranipet (India).[93] Russia leads the list with three of the world's 10 most polluted cities. Linfen, China, was put on the list to represent the numerous Chinese cities that are choking on their industrial air pollution, and Ranipet, India, is an example of serious groundwater pollution by heavy metals. Kabwe, the second largest city in Zambia, is home to 250,000 residents, but decades of unregulated mining and smelting operations have poisoned the soil and water sources. Haina in the Dominican Republic has severe lead contamination, the result of battery recycling, a problem common in the developing world. In some cases the population of these cities number only several hundred thousands, yet the concentrations of pollutants pose the most extreme threats to human health and are a major source of death, illness and long-term environmental damage.

There are two future directions for megacities of the twenty-first century. They could become places where new policies and politics will reorganize these spaces to make them more livable. Or, they could become dysfunctional, non-livable settlements that will cause severe local, regional and global environmental degradation.

Postindustrial cities

In the wake of deindustrialization of many cities in the West there is a reorientation of the relationship with the physical environment. This takes a variety of forms, including the reimagining of the city and its environment, the production of postindustrial landscapes and the cleaning up of toxic sites. In this section we will consider elements of all three.

The city of Syracuse, New York, is a good example of the reimagining of the industrial city. This city of approximately 145,000 people in upstate New York has an industrial history based on salt production and later on a range of manufacturing and metal-based production. A major effect was the pollution of the local environment especially Onondaga Lake. The city celebrated its industrial base with images of factories and salt fields. In 1972 the mayor of the city organized a design competition to replace the one hundred-year-old seal. There was community resistance, and it was only in 1986 that another mayor was able to introduce a new city logo. This logo represents a clean lake and an urban skyline with not a factory chimney to be seen (see Figures 4.1 and 4.2)

Figure 4.1 *Syracuse Logo, circa 1848–1987. This logo celebrates the industrial city.* Source: Photo by John Rennie Short

Figure 4.2 *Syracuse Logo, 1987–present. This image represents the post-industrial city.*
Source: Photo by John Rennie Short

Industrial cities in the developed world have a difficult time in an era of world competition and of the global shift of industry toward much lower cost centers. To be seen as industrial is to be associated with the old, the polluted and the out-of-date. Cities such as Manchester in the UK, Syracuse, Pittsburgh and Milwaukee in the United States, and Wollongong in Australia all have been (re)-presented in more attractive packages that emphasize the new rather than the old, the fashionable postmodern rather than the merely modern, the postindustrial rather than the industrial, consumption rather than production, spectacle and fun rather than pollution and work.

Waterfront redevelopment

The reorientation is evident in reconstructions as well as representations. The process of deindustrialization sometimes allows opportunities for urban redevelopment as factories are abandoned and new geographies of production and circulation leave old docks and railway lines economically redundant. Beginning in the 1960s, continuing in the 1970s and accelerating in the 1980s, waterfront redevelopment became a widespread process in the production of new urban landscapes. The processes of de-industrialization left many cities with abandoned warehouses and buildings, and unused port facilities on their waterfronts. The advent of port containerization meant many older port facilities were inadequate for the new technology and became obsolete. As a result, urban waterfront redevelopment projects were among the most prominent examples of urban renewal in the 1970s and 1980s.[94] Cities in the US, Canada, Europe, Australia, New Zealand and Japan have transformed their waterfronts into vibrant, public spaces that attract locals and tourists. In London, the Docklands were transformed into a vast complex of multi-use spaces that include office buildings, shops, museums and residences. This project stretched from the City through the East End, some 5,500 acres. Figure 4.3 shows construction along the waterfront and Figure 4.4 shows a completed residential project. In Boston and Pittsburgh waterfronts became the new festival spaces filled with sports stadiums, restaurants and hotels. Baltimore's Inner Harbor is often cited as a model US waterfront redevelopment project. It has become the city's gathering place: home to the national aquarium, two sports stadiums, hotels, restaurants, museums, high-rise condominiums and hotels (Figure 4.5). In Barcelona the urban renewal associated with the 1992 Olympic Games involved the opening up of the old docks to a harbor waterfront with a pedestrianized walkway. Until the mid 1980s, the city had turned its back to the sea, leaving only warehouses and docks at the water's edge. After the makeover the harbor front became a congenial place for leisure consumption rather than for production and storage. Waterfront development is not limited to cities in the

Figure 4.3 Construction at London Docklands
Source: Photo by John Rennie Short

Figure 4.4 Residential housing at London Docklands
Source: Photo by John Rennie Short

Figure 4.5 *Baltimore's Inner Harbor*
Source: Photo by John Rennie Short

rich world. In Beijing, China, waterfront development has occurred alongside rapid economic growth (see Figure 4.6).

Such large-scale development as in Docklands or Baltimore is not without its costs. The reconstruction of Baltimore's Inner Harbor cost $2.9 billion. But there also social costs. The diversion of funds to Baltimore's Inner Harbor contrasts with the city's poor public school system and the perceived decline in many public services. In some cases waterfront development is part of a valorization of selected parts of the urban landscape, which allows for the further enrichment of real estate interests, sometimes at the expense of social welfare programs. While Baltimore's Inner Harbor flourishes many inner city neighborhoods continue to experience high crime rates, population loss and housing abandonment.

Superfund sites and brownfields

Industrial manufacturing processes in urban areas often generate hazardous waste. For much of the twentieth century, hazardous wastes were largely unregulated. During the 1970s and 1980s many cities of North America and Europe saw a significant decline in industrial manufacturing. Many industrial manufacturers

Figure 4.6 *Waterfront development in Beijing*
Source: Photo by Michele A. Judd

relocated to countries where labor was less expensive or environmental regulations were more lax, abandoning their factories and leaving behind a legacy of pollution and environmental contamination.

In the US, the 1980 Comprehensive Environmental Response Compensation and Liability Act (CERCLA), more commonly known as the Superfund, established a multi-billion dollar federal trust fund to pay for emergency measures as well as cleanup of sites for which responsible parties could not be identified. Superfund sites are found everywhere, including rural areas. Many superfund sites are in and around urban areas, where there has been a long history of water or land pollution. The EPA has identified a National Priority List of some 30,000 sites they consider to be the most immediate hazards and to which they prioritize funds for cleanup. A full list of current Superfund sites can be found at http://www.epa.gov/superfund/index.htm The Superfund legislation set an important precedent: polluters were responsible for cleanup, and in cases where polluters were no longer in business or there was no identifiable responsible party, the federal government would assume the costs of cleanup.

In addition, 1986 amendments to Superfund established a Community Right to Know Act that requires industry to report publicly their use and releases of more than 100 toxic chemicals. That information must be made available at County Boards of Health as well as public libraries. This amendment not only provides incentives for pollution reduction, it also adds the element of the fear of public disclosure for those companies not in compliance with regulatory standards.

By the mid 1990s many derelict site areas had not been cleaned up or were a low priority on the Superfund list. A new term emerged to describe these contaminated areas: brownfields. Although brownfields can be found anywhere, many are concentrated in cities that have hosted an array of manufacturing activities. Brownfields share numerous characteristics outlined in Table 4.8. One widely accepted definition of brownfields is the one provided by the US EPA, which defines them as "real property, the expansion, redevelopment, or reuse of which may be complicated by the presence or potential presence of a hazardous substance, pollutant, or contaminant".[95]

The EPA estimates there are more than 450,000 brownfields in the US. Brownfields, unmediated (not cleaned up) can have health impacts that include cancer mortality, reproductive effects and chronic disease morbidity. In order to promote the cleanup and redevelopment of these urban sites, the EPA has developed the Brownfields Initiative, a successor of sorts to the Superfund. It has encouraged many states and local governments to establish programs that provide tax credits and other financial incentives such as low-interest loans to attract private investment. There are grants that provide funding for brownfield inventories, environmental assessments and community outreach. There are also loans that provide funds to clean up these sites and job training grants that provide training for residents of brownfield communities. Many loans and grants prioritize brownfields in low-income areas or economically depressed areas.

Some brownfields are large and require extensive rehabilitation and decontamination work, and the costs are considerable. Prospective buyers and developers of contaminated properties are often leery of becoming responsible for cleanup and many of the parties who caused the original contamination may be long gone. As a result federal and state governments have attempted to reduce the barriers to redevelopment and to financially support projects through grants, low-interest loans and other subsidies.

The US EPA created a grant program to assist cities redevelop brownfields. Between 1995–2004 the EPA invested over $700 million in brownfields, providing

Table 4.8 *Characteristics of brownfields*

Industrial heritage
Integral part of the urban structure
Real or perceived contamination problems
High local unemployment
Decline of municipal revenues
Adverse affects on urban life
Need for intervention from outside to be brought back to beneficial use
Offer development without encroaching on existing green spaces

Source: Adapted from US Environmental Protection Agency

grants to states, local governments and nonprofits.[96] The grants are used to help plan the decontamination and the redevelopment of these spaces. In addition, the EPA has tried to scale back environmental requirements that would be perceived as too onerous for development. It may not be necessary to have the same cleanup requirements at a parcel slated to be an industrial reuse area compared to a children's day care facility. State and federal governments have tried to tailor cleanup requirements to meet the expected future use of the properties.

One example of a successful public–private brownfield venture is the Highland Marine Terminal in the heart of southeast Baltimore's industrial port district.[97] For nearly a century, the site was a copper-processing plant. The site was contaminated with lead, arsenic, beryllium, copper, nickel, silver and selenium. As one of the first of Baltimore's brownfield projects, plans were to turn the site into warehouse space in a part of the city that had insufficient warehouse space. The redevelopment deal consisted of a state-subsidized loan, loan guarantees and $40,000 in grants from the city and state. The state of Maryland also put in another $1 million in low-interest loans. The developer razed some 100,000 square feet of buildings and cleaned up the site. The project built or rehabilitated more than 700,000 square feet of warehouse space and seven acres of outside storage, all of which is currently fully leased. Developers were quick to state that without the brownfield program, environmental issues might have deterred them from undertaking the project. Several decades of successful cleanups and redevelopment projects show that given incentives, developers find it profitable to absorb the risks of purchasing, cleaning and reusing contaminated land.

Cities in Europe have also been affected by brownfields. The amount of land defined as brownfields is difficult to assess. In Germany there may be as many as 300,000 sites covering about 128,000 hectares; in France approximately 200,000 sites; in the UK some 100,000 sites; and in Belgium some 50,000 sites covering at least 9,000 hectares.[98] Environmental contamination not only has an adverse effect on the environment, but is also considered a significant barrier to the economic and social redevelopment of the city. In addition, there is growing evidence to suggest that many urban brownfields are situated in areas with higher concentrations of minority populations and households below the poverty level, raising important questions about environmental justice.

Brownfields are redeveloped in a variety of ways: for industrial reuse, commercial or residential uses, and also as green spaces such as parks, playgrounds, trails and greenways. The projects can be modest – the reuse of a single isolated property – or more ambitious such as the revitalization of an entire distressed neighborhood. In the US, there has been a tendency to redevelop brownfields into spaces that have immediate economic benefits, and cities have focused more on commercial or residential uses. In European and Canadian cities, however, there has been more emphasis on converting derelict brownfields into green spaces. For example, between 1988 and 1993 19 percent of brownfields in the UK were converted into green spaces.[99]

In Derbyshire, a former coal-mining facility is being transformed into new deciduous woodlands (mainly willow trees). The willow trees will be harvested and used at the nearby tip as a renewable energy resource to heat boilers for nearby commercial buildings.

Geographer Christopher De Sousa has examined the transformation of several brownfields into green spaces in the city of Toronto.[100] He found that many of the larger brownfield-to-green space projects involved sites that were near or within existing parklands or floodplain areas. The projects reintroduced native trees, shrubs, wildflowers and plants that enhanced the ecological integrity of the area and improved flood protection, storm-water control and offered increased recreational opportunities for many under-serviced communities. De Sousa notes, however, that all the projects were carried out by the public sector, as private interests were deterred by the perceived costs and lack of benefits.

Pauline Deutz looked at what she termed eco-industrial parks in North America and found that they were often used as a form of place promotion with economic development issues sometime trumping ecological objectives.[101]

Brownfield redevelopment is a form of land recycling that can restore and regenerate formerly derelict and toxic urban spaces. Brownfield initiatives have been integrated with community economic redevelopment and job creation, but have also been important aids in health and safety issues, neighborhood restoration and the reuse of urban space to counter suburban sprawl. In formerly industrial cities, reintegrating brownfields into urban space promotes sustainable development.

Sprawl

Recent metropolitan growth throughout the world, but especially in the developed world, is characterized by a low density spread across the landscape. Suburbanization of jobs and residences has extended the urban region further out from central cities. We will use the term sprawl to refer to low density, car-dependent development with a low land use mix, and relatively low connectivity. The term *sprawl* is not just a technical term. It is often used as a negative connotation as many critics use the contemporary suburban landscape to marshal a critique of society. For many, sprawl is a text that tells a story of environmental degradation, social fracturing and loss of community.[102] There are also more recent defenders of sprawl. Robert Breugmann provides a vigorous defense of low-density scattered urban development at the edge of cities. His book is an extended argument that sprawl has always been a part of urban development, and that it reflects the desires of the consumers. The criticisms of sprawl, he believes, reflect the class bias of intellectual elites.[103]

While perhaps easy to use, sprawl is more difficult to operationalize. Recent, detailed work on measuring sprawl reveals a complex picture. Wolman et al. tackle the difficult question of how to define and measure sprawl. They show that the measurement depends on which land area forms the basis of analysis. Even within one country, sprawl takes different forms. In the dry Sunbelt of the US, for example, sprawl is of higher density than in the rest of the country because of aridity, public lands and slope constraints.[104]

The environmental impacts of sprawl have long been noted and include loss of open space, greater air pollution, higher energy consumption, increased risk of flooding, ecosystem fragmentation and reductions in diversity and species.[105]

The loss of green space is particularly acute in areas of rapid growth. The Washington, DC, metro area lost 50 percent of its green space between 1986 and 2000. In this short time period the percentage of developed land increased from 12.2 to 17.8. As sprawl continues outwards, more land is needed. The rapid growth in Montgomery County just north of Washington, DC, has resulted in increased traffic. A new inter-county connector, first proposed in the 1970s, is now scheduled for construction. The 18-mile, $2.4 billion road linking Interstates 270 and 95 will impact streams, wetlands and other sensitive ecological areas in its path. Pressure from community and environmental groups has resulted in the construction of bridges over thewetlands and floodplains but the impacts will still be large. And in the wake of the road will come more sprawl. Ironically, the existing solutions to sprawl often create the conditions for greater sprawl.

Sprawl is a form of development that is very often too diffuse to support public transport or easy walking. The heavy and in some cases total reliance on private auto transport imposes a heavy environmental price in terms of air pollution, and the increasing dedication of space for roads and parking. The reliance of a built form precariously balanced on one fossil fuel with large and fluctuating costs raises issues of long-term sustainability. Matthew Kahn showed that the US suburban households drive 35 percent more and use more than twice as much land than their urban counterparts. However, he also demonstrated that the greater vehicle use did not lead to greater air degradation because of stricter emission controls. In other words technology and regulations can mitigate against the environmental impacts of sprawl.[106]

Urban sprawl creates more paved surfaces. When 10 percent to 15 percent of land surface is paved then increased sediment and chemical pollutants reduce water quality; at 15 percent to 20 percent there is markedly reduced oxygen levels in streams; and at 25 percent many organisms die. Many studies decisively document the impairment of streams with the increase of urban land use.[107]

The issue of species depletion and ecosystem fragmentation is more complex. If the land use change is from a single land use such as green pasture or wheat fields to suburbs, then a richer and wider variety of ecosystem may be produced with public and private gardens providing a wider range of diversity than a simple monoculture. New bird and animal species, for example, adapt to suburban gardens. There is need for a more careful assessment of the environmental consequences of land use changes at the city's edge. If we exchange a field of corn, genetically engineered, with yields only sustained by chemical infusions, for suburban garden plots, is this, by definition, a loss of environmental quality? Much of the debate about suburbanization and urban sprawl has been conducted in the context of an anti-suburban rhetoric. More detailed case studies may provide a more reasoned account.

The new industrial city

In the wake of a global shift in manufacturing employment, new industrial cities have developed in developing countries. We can follow the wave of the industrialization of cities from Japan and Korea and in the 1960s and 1980s, through to coastal China in the current era. The newest industrial cities often emerge in an intense period of rapid economic and population growth coupled with few environmental regulations or little enforcement of environmental regulations. The result is unprecedented levels of air, land and water pollution. In many ways, the new industrial city resembles the industrial city of the nineteenth century. In some cases, however, unprecedented levels of urban pollution are generated.

The rise of China as a new global economic force has been unrivaled in recent years. GDP has quadrupled and per capita incomes have risen 3-fold in many cities. Since 1979, China has implemented economic reforms that transformed selected cities, designated "free-market zones" that encouraged foreign investment. These zones include the cities of Shanghai, Beijing and Guangzhou. Two decades of unparalleled economic growth, swelling urban populations and often unchecked emissions from automobiles, factories, domestic heating, cooking and refuse burning have made China's cities prone to air pollution.

Consider the capital city of Beijing, which has been rapidly transformed in the past 20 years. Foreign investment, focused on export processing, retail and insurance, has generated new levels of affluence; in addition, the government has attracted industry with development zones that concentrate the location of certain types of economic functions. For example, in the southeast of Beijing, the smaller city of Tianjin has become a major international port city; to the east Tangshan is now a major center of heavy industry and coal mining. Forming a ring around Beijing is a series of major industrial areas including textile mills, iron and steelworks, machine shops, chemical plants, and factories manufacturing heavy machinery and electronic equipment. The city downtown has large commercial/financial districts, busy shopping areas, and thousands of new stores and restaurants. Growing per capita income has generated a construction boom of residences and commercial properties.

But at the same time there has been unparalleled environmental pollution. In 2005, Beijing was listed as one of the ten worst cities in the world for both air and water pollution. As in most Chinese cities, soft coal is the predominant form of energy. The heavy reliance on coal means enormous amounts of sulfur and nitrogen oxides are released. Respiratory diseases have become one of the country's biggest health risks. The World Bank estimates that 10,000 people die prematurely each year from particulate pollution in just four of China's largest cities: Chongqing, Beijing, Shanghai and Shenyang.[108] Beijing and the surrounding north-east China region have the world's worst nitrogen dioxide pollution.

Shanghai was designated a special economic region under the 1979 reforms. Under the reforms foreigners and foreign investment began to flow into the city generating economic growth rates of about 7 percent a year. A zone on the east side of the old city core, Pudong, has attracted high-tech industries and financial services. Throughout the 1990s, Shanghai's annual economic growth rate was above 10 percent.

All households in Shanghai have access to piped water, electricity and other services. But, as with Beijing, Shanghai relies on soft coal as a source of fuel for both industrial energy and residential heating. It too suffers from significant air pollution. In addition, some 4 million cubic meters of untreated human waste enter the Huangpu River each day. Currently, the city is home to a significant heavy industry sector, including machinery manufacturing, textiles and steel. The toll of unrestrained economic growth in tandem with unimplemented environmental regulation has created a city in desperate need of pollution control.

In many ways the recent rise of the new industrial cities resembles the Coketowns of the industrial revolution of the eighteenth and nineteenth century. Pollution has blotted out the sun and sanitation systems cannot keep pace with population growth. These new global industrial cities are in need of sweeping and comprehensive air, land and water reform measures.

Environmental factors also place limits on economic growth. Of China's 600 cities, 400 of them have water shortages and approximately 100 have severe water shortages. This puts brakes on continued economic growth. One solution, involving piping water from the south of the country to the north, can only be implemented with heavy costs involving environmental disruptions and the forced relocation of over 300,000 people. Breakneck economic growth with few environmental controls creates not only environmental pollution but also ecological bottlenecks to sustained growth.

Shantytowns

The term "shantytown" (also called squatter housing) refers to unplanned, illegal, informal housing.[109] The housing is considered illegal because the occupiers hold no title to the land, do not pay taxes, and have constructed some form of shelter that does not meet building codes. Because these are illegal structures, they often lack government provisions such as infrastructure or services such as clean water. The World Health Organization estimated that in 2000 more than 600 million urban dwellers in developing cities had no access to clean water, sanitation or drainage (See Figure 4.7). Those who live in shanty settlements face insecure land tenure, exposure to hazards and often lack political voice. Table 4.9 shows the percentage of residents living in cities in substandard or illegal housing. In Mexico City, for example, most of the urban poor live in the shanties on the periphery, where they build their own homes on appropriated land using whatever materials they can find.[110] Mexico City is home to one of the largest shanty cities in the world – Ciudad Netzahualcóyotl – with more than one million urban poor who live in self-built dwellings.

In shantytowns the poorest urban dwellers are exposed to a myriad of environmental and social problems that include:

• Lack of infrastructure providing water, sewage, electricity, trash collection

Figure 4.7 Nairobi informal housing
Source: Photo by David Rain

Table 4.9 Substandard housing in selected cities, 2000

	% substandard housing
Accra	35
Dar es Salaam	60
Delhi	51
Guatemala City	64
Kigali	93
Kinshasa	60
Lagos	15
Lahore	85
Lima	60
Mumbai	47
Nairobi	75
Rio de Janeiro	25
Tunis	25

Source: Global Urban Indicators Database
http:www.urbanobservatory.org/indicators/database

- Disease-causing agents (pathogens) in air, food, water, or soil that impact human health
- Pollutants in air, food, and water that impact human health in both the short term and long term
- Congestion
- Physical hazards such as accidental fires, floods, mudslides or landslides.

In the rest of this section we will examine these items in more detail.

The lack of infrastructure is widespread in cities in the developing world. Consider the following statistics: in Bangkok, 33 percent of the population do not have access to a clean water supply; in Calcutta, 5 million are without clean drinking water. In Khartoum, 95 percent of the population has no municipal sewage system; in many cities in India more than a third of the population have no latrine and must rely on buckets to remove human waste. In Bogotá, more than 2,500 tons of trash goes uncollected each day. On July 11, 2000, the collapse of a rubbish dump in Payatas, Manila, killed 218 people living in shanties at the bottom of the site and left another 300 people missing, trapped under the rotting garbage. "The tragedy of their burial underneath the trash of a world city, off its edge and in the darkness of night, symbolizes the invisible, daily plight of innumerable poor people."[111]

It is estimated that hundreds of millions of urban dwellers do not have piped water supplies and thus have no alternative but to use contaminated water, or water whose quality is not guaranteed.[112] This is not to say that they have no access to water; they do. In some shanties there are public stand posts or public fountains from which residents can fill buckets and other containers. However, there is a significant amount of time needed to obtain water, and often long distances must be traveled to collect it. In addition, there are water vendors (usually private firms) who sell water to the poor, but they often charge 5–10 times as much as the rates for water delivered by a public water system. Many of the urban poor cannot afford private water sources, and hence have an inadequate supply. As a result, many shanty residents use whatever water they can find; often it is contaminated. Water-borne diseases take a tremendous toll on human health. Diarrheal diseases affect an estimated 700 million people each year and account for most water-related infant and child deaths. In addition studies have shown that a high proportion of the shanty residents have intestinal worms that cause severe pain and malnutrition.[113]

The World Health Organization noted that an infant is 40–50 times more likely to die in a shantytown than in a city in North America or Europe.[114] In Delhi, for example, shanty settlements average 300,000 people per square kilometer. With population density at such an extreme, and with so many exposed to untreated sewage and wastes, many shantytowns are places where diseases such as hepatitis and tuberculosis flourish. While most industrialized cities long ago eliminated diseases such as cholera, typhoid and dysentery, these diseases still plague developing cities. In January 1991, an outbreak of cholera was first observed in Lima, then spread to cities and towns in Colombia and Ecuador. By the end of the year it had crossed the continent and more than 350,000 cases had been reported. Among shantytown residents, neonatal deaths are two times higher, mortality from respiratory disease six times higher and mortality from septicemia eight times higher than among the middle class or wealthy in that same city.[115]

Poverty often means that infants and children do not always receive their vaccines for measles, whooping cough and diphtheria, exposing these children to diseases that more wealthy populations no longer contend with. Malnutrition in shantytowns tends to be higher than in either cities or rural areas.

A particular type of pollutant unique to populations living in shanty cities is indoor air pollution that comes from indoor smoke, a result of cooking over open wood or dung fires. Most shanties lack fans or exhaust systems, and because they often have no electricity or gas, cooking takes place over open fires.

The health impacts of consistent exposure to smoke and fumes has been underestimated and understudied, but we do know that burning coal, wood or other biomass fuels can cause serious respiratory and eye problems.[116] Research has shown that concentrations of total suspended particulates are 10 to 100 times higher in indoor dwellings. Chronic effects of exposure include inflammation of the respiratory tract, which in turn increases vulnerability to acute respiratory infections such as asthma, bronchitis and pneumonia. Women are often heavily exposed because they spend several hours a day at the stove; infants and young children may also be heavily exposed since they remain close by their mothers. The exposure of infants and young children to indoor air pollution, combined with malnutrition (often common) leads to a greater prevalence of chronic bronchitis.

Another pollutant shanty dwellers must deal with is the removal and safe disposal of excreta and waste water. While some cities have concentrated on improving the provision of clean water, fewer have dealt with issues of sanitation. Many of the waterborne diseases, however, are excreta-related, such as schistosomiasis, hookworm and tapeworms. Shanty dwellers dispose of excreta in pit latrines, bucket latrines, or "flying toilets". Flying toilets refer to defecation that has been placed in plastic bags, which are then thrown into ditches, gullies, streams, canals and rivers, where the human waste remains untreated and may further contaminate water supplies down river. Figure 4.8 depicts the "flying toilet" in Mahare, a shanty in Nairobi, Kenya.

High density and congestion are hallmarks of shanties. In the city of Calcutta, there are some 1,500 shanty colonies. The average dwelling measures 6 foot by 8 foot, yet houses 6–8 people. Such high population density facilitates the spread of contact diseases such as influenza, meningitis and TB as well as food-borne diseases. The frequency of contact, the density of the population, and the concentration and proximity of infectious and susceptible people in a urban population promotes the transmission of diseases.[117]

The urban design of shanty settlements exacerbates risks. The dense settlements mean that narrow paths or streets (usually unpaved) restrict vehicular access. As a result, ambulances, fire-fighters and police often have trouble accessing areas. There is also the lack of recreational space or safe places for children to play.

The urban explosion of the past fifty years has forced expansion onto new spaces on more vulnerable sites such as steep hillsides, flood-plains or in areas with unstable soil conditions. Many shanty structures are often erected on such marginal lands because standard, legal housing long ago claimed the best, most

Figure 4.8 *"Flying toilets" in an open sewer in a Nairobi slum*
Source: Photo by David Rain

secure land in the city. The poor are often forced to settle on land subject to higher environmental risks of landslides and flooding.

Shanty cities are very vulnerable to disasters such as landslides, fire and erosion. Those living in informal settlements have a higher probability of being killed or injured in an accidental fire, earthquake, or flood or during a storm than people in the formal housing sector of a particular city. Fires are common in shanty dwellings because the building materials tend to be inflammable and most of the cooking takes place over open stoves or fires. The lack of electricity means people heat and light their dwellings with candles, or kerosene lamps. A study of injuries in shanties in Karachi found that most of the injuries were due to falls, burns and cuts.[118] Most of the burns were suffered by women who were cooking and young children who accidentally burned or scalded themselves.

In Mumbai, a hillside gave way and the informal shanty homes perched precariously on it were swept away. In this instance, heavy rain contributed to the landslides, but the "real causes" were due to the fact that low-income groups could find no land site that was safe and to the failure of government to ensure a safer site or to take measures to make existing sites safer.[119]

The impacts of shanty life disproportionably impact women and children. In most shanties, infants, children and women bear the heaviest burden of air- and water-polluted sanitation and vulnerability to environmental hazards.

Shanty towns can provide a platform for rural urban migrants to the city, but at some cost. The dichotomy between slums of hope and slums of despair summarizes the benefits and costs. The environmental costs can create a web of multiple deprivation that traps the poorest. The shantytown conditions can exacerbate the risk to human health and the vulnerability of these residents to other problems. For example, the lack of basic infrastructure increases the likelihood of pathogens spreading diseases. The lack of water services means that fires are more difficult to contain and put out, and the lack of paved roads poses difficulties for emergency and rescue services. Many of the urban poor in developing cities have multiple deprivations.

Rapid urban growth in the developing world need not produce either the heavily polluted industrial city or the particularly vulnerable shantytown. Some cities such as Curitiba in Brazil have mitigated environmental degradation. And there have been efforts to help improve the social and environmental quality of the urban poor. For example, in Belo Horizonte, Brazil, a PROFAVELA project helped squatters obtain land tenure, and thus, connection to municipal service networks. Some shanties in Lima, Peru, have organized trash pickup teams, which pedal tricycle-like carts along fixed routes in the slums. In Chapter 12 we document in greater detail the efforts to provide a more ecologically sustainable relationship in the urban environment.

Guide to Further Reading

On megacities and shantytowns

Hardoy, J., Mitlin, D. and Satterthwaite, D. (2001) *Environmental Problems in an Urbanizing World*, London and Sterling, VA: Earthscan.

Hardoy, J. and Satterthwaite, D. (1989) *Squatter Citizen: Life in the Urban Third World*, London and Sterling, VA: Earthscan.

Economy, E. (2004) *The River Runs Black: The Environmental Challenge to China's Future*, Ithaca, NY: Cornell University Press.

Rock, M. (2002) *Pollution Control in East Asia: Lessons from Newly Industrializing Economies*, Washington, DC: RFF Press.

On sprawl

Frumkin, H., Frank, L. and Jackson, R. (2004) *Urban Sprawl and Public Health*, Washington, DC: Island Press.

Johnson, M. P. (2001) "Environmental impacts of urban sprawl: a survey of the literature and a research agenda", *Environment and Planning A* 33: 717–735.

Kahn, M. E. (2000) "The environmental impact of suburbanization", *Journal of Policy Analysis and Management* 19: 569–586.

Saunders, W. S. (ed.) (2005) *Sprawl and Suburbia*, Minneapolis and London: University of Minnesota Press.

Squires, G. D. (ed.) (2002) *Urban Sprawl: Causes, Consequences and Policy Responses*, Washington, DC: Urban Institute.

On urban regions

Short, J.R. (2007) *Liquid City: Megalopolis and the contemporary Northeast,* Washington, DC: Resources For the Future.

On urban trends

Hall, T., Hubbard, P, and Short, J.R.(eds.) (2007) *The Sage Urban Compendium,* Londen: Sage.

Scott, A.J.(ed.) (2004) *Global City-Regions: Trends, Theory and Policy,* New York: Oxford University press.

Part II
Urban Environmental Issues

5 Urban sites

We can make a distinction between location and site. Location refers to relative space, the space of connections and the space of urban hierarchies, economic transactions and social relations. It is a space abstracted from territory. Site, in contrast, refers to the absolute space that a city occupies. In much of recent urban geography and indeed of urban studies in general, the abstract space of location is a more dominant theme, an intellectual trajectory that tends to ignore, marginalize or simply forget the importance of absolute space to understanding cities.

But there is also a large and growing body of work that looks explicitly at the place of cities. These studies range in theoretical orientation from the landscape analysis school, through political ecology to critical social theory. New Orleans, for example, has been the subject of a number of works. The cultural geographer Pierce Lewis looks at the making of the urban landscape while Craig Colten, from a more political ecology perspective, tells how the city was shaped in the attempt to conquer nature.[120] Reyner Banham also used the physical geography of a city in his evocative and classic conceptualization of Los Angeles. He outlines the place and role of what he terms: surfurbia, foothills, plains of id and autopia in his part social, part physical model of the city.[121] In his 2003 book, *Paris: Capital of Modernity*, especially in part 2 *Materializations*, David Harvey adopts a more critical perspective in examining the relationships between an urbanized nature and emerging social relations.[122] And in his study of Mumbai, Arjun Appadurai paints a detailed picture of the urban landscape and its connection with cash and capital. He describes a sense of an "immense landscape of street-level traffic" and shows how urban cleansing was invoked in ethnic sectarianism.[123] While these and other studies tend to be very focused on the local, the ideographic and the unique, collectively they remind us of the role of place in the making of the cities, how the urban transformations of place affect both physical and social relationships. Cities occupy sites and this occupancy creates constraints and opportunities, provides the site of material culture as well as the imaginings of urban dreams and fantasies. The site of cities is the basis for economic transactions, the place of social relations and the setting for the production of urban images.

While the dominant theme of urban conceptualization is a point in a two-dimensional surface, much is gained from a three-dimensional view. Meyer, for example, highlights the role of altitude in the early development of the US city. In some cities the high ground was avoided because it was difficult to pump water or to get horse-drawn fire services so it was more of a fire hazard. The early forms of public transport also found steep hills hard going. Later, as altitude was more easily overcome, the high ground became the site of more elite residences away from the pollution and dirt of the lower ground.[124] In Los Angeles, as Banham noted, there is a clear link between altitude and socio-economic status with the Hollywood Hills and the plains of Crenshaw providing polar examples.

Altitude also plays a more general role in the (literal) life of a city. Mexico City, for example, is surrounded by endemic dengue fever but the city has remained relatively free of the disease because of its altitude of 2,485 meters. The fever is contracted through the bite of an infected Andes mosquito. This type of mosquito does not survive at higher altitudes and so dengue fever is more prevalent at lower altitudes. In Venezuela, for example, most of the cities are at a much lower altitude and dengue fever poses a greater threat.[125]

Adding the third dimension provides a greater understanding of cities. But more than just an understanding of altitude is needed to obtain a multi-dimensional view of cities. We have to understand the environmental milieu that surrounds, encapsulates and structures a city. Consider just one type of environmental context - rivers. Many cities are situated beside rivers; they grew up beside specific rivers and the history of the city is, in large part, a tale of its unfolding and changing relationships with the river. To understand Glasgow, London, Florence, Cairo, Phnom Penh, Dublin or Calcutta it is necessary to understand the Clyde, the Thames, the Arno, the Nile, the Mekong, the Liffey and the Hooghly. The evolution of London, for example, can only be understood through its changing relationship with the Thames; and not only for specific former dockside areas such as Canary Wharf, but also to explain the social geography of areas north and south of the river. In Southeast Asia, Milton Osborne looks at the relationship of the Mekong to cities along its source.[126] And even in cities where the river does not immediately spring to mind or is not visible in the landscape, the urban–river connection is there. Blake Gumprecht tells the story of the Los Angeles River. Before western invasion the river system supported a dense network of Native American settlements. Later it proved the main reason for the location of a new city. The river provided drinking water and irrigation for farms. Orange groves were watered with river water. As urbanization extended along the riverbank the seasonal floods caused more and more damage. A series of winter storm floods in the late nineteenth and early twentieth centuries led to calls for greater river management. The river was channeled into 51 miles of concrete culverts. Now it is a completely urban river occasionally running through an entirely human-made channel system, little more than a small trickle of water flowing through a wide concrete scar through the city. Gumprecht's book is subtitled *The Life, Death and Possible Rebirth*. He suggests that the exhumation of the river and the greening of its riverbank could become important goals of ecological restoration, environmental improvement and urban renewal.[127] Around the world, cities in developed countries are reassessing and revaluating their river connections, often moving from a narrow economic resource dependency and dumping ground mentality to a sense of a more sustainable recreational mixed-use set of options. We can tell much about a city from the nexus of its river connections.

In the rest of this chapter we will approach this broad theme of city sites with a series of theorized case studies of particular types of city–environment relations. It is important to remember that the environment structures but does not mechanically determine a city's development.

On the beach

One of the most iconic images in the photographic history of Australia is Max Dupain's 1937 "Sunbaker". It shows the head and shoulders of a young man on the beach. His head lays on his forearms, the short shadows suggest an overhead summer sun. Max Dupain (1911–92) was one of Australia's most celebrated photographers. He photographed many Sydney scenes including streetscapes, beaches and buildings all in a distinctly modern style. By the 1970s his images were recognized as capturing the essence of Australia. But among the thousands of his photographs it is the black and white one of a man sunning on a beach that has become the most widely recognized. The icon of the man on the beach entered the national imaginary as well as international associations of Australia and Sydney.

The earliest settlers who inhabited the area where the city of Sydney now stands were the Eora people. The coming of the First Fleet in 1788 rudely disrupted this coastal people's world of fishing, farming and hunting-gathering. Botany Bay and then Sydney Cove became a far-flung outpost of an imperial Britain.

The city grew to become the largest in the nation and by the late twentieth century was the global gateway of Australia.

Sydney grew from the Inner Harbour in all directions both inland and along the coast (see Figures 5.1 and 5.2). The result was an incorporation of beachscapes into the urban fabric. There are 38 beaches along the sea coast that range in size from the expansive 5 kilometers of Cronulla to the pocket-sized 70-meter strand of Clovelly. They are strung out along the coast from Palm Beach in the north to south of Botany Bay. There are also 32 beaches inside the Harbour.

The ocean beaches are particularly striking because they have beautiful golden sands, clean ocean water, high swells that provide photogenic waves and a geology of sandstone headlands that surround the beaches in a graceful crescent arc. The Harbour and ocean beaches are an integral part of the urban fabric, adding light and space and a closer connection to the sea. Sydney's reputation for a free and easy attitude may in part be due to its less confining urban space. In contrast to the sense of confinement in built-up inland cities,

Figure 5.1 *Sydney, an aerial view*
Source: Photo by John Rennie Short

Figure 5.2 *Sydney: waterfront housing and development*
Source: Photo by John Rennie Short

Sydney's expansive relationship to the ocean and the open water creates feelings of space and openness.

Sydney and the beach

"The beach is Australia's true democracy ... we Australians did not derive our freedom from bewigged Georgian founding fathers and their tablets of good intentions ... we have found our freedom by taking our clothes off and doing nothing of significance, and by over the years refining and elevating this state of idleness to a culture now regarded highly in the world's most fashionable places ... Whatever racists and Jeremiahs may say, Australia, a society with a deeply racist past, has absorbed dozens of diverse cultures peacefully. The beach and the way of life it represents are central to this. Today the sons and daughters of these people are often the majority on Bondi Beach, where lifesavers have Italian, Greek and Turkish names and board riders are Vietnamese ... the beach is theirs now."

Source: John Pilger (1989) *A Secret Country*, London: Jonathan Cape, pp. 10–12.

The beaches are also an integral element of life in the city. The English emigrant Henry Gilbert Smith turned the fishing village of Manly into an ocean resort in 1857. Bathing was initially regulated. A law that allowed bathing only between eight in the evening and six in the morning was only relaxed in the early twentieth century. The editor of a local newspaper first publicly defied the ban in 1902 and police refused to prosecute. People soon came to the beach to swim and sunbathe and hang out. Swimming clubs, life-saving clubs and later surfing clubs provided the social binding that connected people in what could truly be called an urban beach culture. The male lifesaver became an image of Australian masculinity, surfing helped to define a youth culture, topless sunbathing was an important element in women's liberation in Australia and in the articulation of changing sexual mores. Sydney was a city where the body on the beach was an important element of the urban experience. Social changes were reflected and refracted through the prism of this sensual beach experience.

There were so many beaches that there was marked differentiation. The northern beaches, located in the suburbs, such as Curl Curl were considered more sedate than the more youthful and cosmopolitan Bondi. Around the larger more popular beaches restaurants and hotels, bars and shops soon developed. The beach and the coastline also shaped the social geography. In Sydney the rich tend to live in the inner city along the waterways and coast; the poor live further inland more distant from the beach. The inland western suburbs are more blue-collar than the inner city Double Bay.

The beach is a contested place. In December 2005 civil unrest broke out along the beaches of Cronulla, south of Botany Bay. The general context was not only the changed world of post 9/11 but also the particular Australian experience of a terrorist bombing in Bali in 2002 that killed 88 Australians. Anti-Islamic feeling was high. The more local context was the location of Cronulla. It is one of the few beaches accessible by rail and hence more accessible to the poorer residents of the western suburbs. The suburb itself is one of the more Anglo-Celtic suburbs of the city, less impacted by the large foreign-born immigration of the city. Because of its easy accessibility from the western suburbs, Cronulla beach had become a favored beach for poor immigrant youths. Lebanese youths had attacked two lifeguards the week before. Anecdotal evidence from local informants suggests that conflict had been caused by competition for the attention of young women on the beach. On Sunday December 11, 2005, a mob of over 5,000 white men marched along the beach attacking people they thought were Lebanese. The mob had been organized by right wing and neo-Nazi groups eager to show that the beaches were only for "real" Australians.

The beach has also played part in the selling of Sydney. The beach and bridge, the sun and Opera House are part of the internationally marketed and promoted images of the city. The success of the city in competition with Melbourne in the race to become the primary global gateway can in part be put down to the more recognizable urban iconography and the role that the tempting beach culture has played in the successful international marketing of the city. Sydney has an ensemble of some of the most internationally recognized urban images, the Harbor Bridge, Opera House and Bondi Beach. The hedonistic pleasures of the beach beckon people from near and far.

Cities in the desert

Mecca and Las Vegas: two unlikely cities to be paired together. One a destination point for religious pilgrims, the other for gambling and entertainment. But both cities are located in a desert and this location has helped to shape the cities. The idea of a pilgrimage, whether for the soul or the body, is enhanced by the sense of a literal or metaphorical journey through the desert to an oasis.

Mecca is situated in the Sirat Mountains, 45 miles from the coast of the Red Sea. The city is dry and hot. Rainfall is less than five inches per year although winter flash floods can sweep through the usually dry streambeds of the Wadi Ibrahim. Temperatures are high and in the summer can reach over 100° F. The city has long been an oasis, wells capture underground water sources and one well, Zam Zam, was considered a gift from God to Abraham and his wife Hagar. An oasis in the middle of harsh desert was interpreted as divine intervention. By the time of the Romans and the Byzantines the city was an important trading center and a center of pilgrimage. Pilgrims came to visit the Kaaba, a 50-foot high cubical building, supposedly built by Abraham and his son Ishmael over 4000 years BCE, housing a black stone that was said to have fallen from God at the time of Adam. The city was the birthplace of the prophet Muhammad around 570. He was forced to flee the city in 622 but returned eight years later, destroyed the pagan idols and declared the city the center of Muslim faith. The city is the most sacred site of Islam and a pivotal point in the life of Muslims. Devout Muslims pray five times a day in the direction of the city and are supposed to make a pilgrimage, the hajj, to the city at least once in their lifetime (see Figure 5.3).

Islam has 1.2 billion adherents and is the fastest growing religion in the world. With the growth of the believers has come the growth of the hajj. In 1950, approximately 250,000 people made a pilgrimage to the holy city; by 2006 the

Figure 5.3
Pilgrims in Mecca
Source:
http://
www.
worldcity-
photos.org
/SaudiAra
bia/ SAU-
mecca-
webshoth
ajj
20012.jpg

Cooling desert cities

Issues of global warming are often presented as topics suitable only for international forums and the coordination of national policies around the world. Increasingly, as some national governments, especially the US, resist global regimes, regions, states and even cities are getting more involved. In 2006, 22 of the world's largest cities made a combined pledge to reduce greenhouse gases by implementing policies and sharing technology. In the US, the more progressive states have set emission targets for greenhouse gases, some requiring power plants to generate a portion of their electricity from renewable resources, and 11 states have followed California's lead in adopting stricter vehicle standards for greenhouse gases than the federal standard.

Even smaller cities can play a part. Keeping cool in the desert involves smart design as well as sophisticated technology. It also involves a range of innovative programs. Alice Springs is situated in the middle of Australia. The semi-arid climate produces hot dry summers and warm winters. A local initiative prompted by the desert knowledge organization called Cool Mob, has created a program to reduce electricity, transport and water use, minimize waste and encourage recycling. The positive environmental outcomes are pursued by the provision of energy-saving information to households, access to a range of discounts, and workshops aimed at retrofitting older housing and building more environmentally sensitive newer housing.

For more information, go to Website http://dka.coolmob.org

numbers had increased to over 2.5 million. In the simplest form of the hajj, pilgrims make their way to the city in the Islamic month of Dhu al-Hijjah, the men wearing only a simple plain garment composed of two white sheets. Recreating Hagar's desperate search for water they walk seven times between the two hills of Safa and Mawah and then circle the Kaaba seven times counterclockwise, the first four at a quick pace, the last three more leisurely. The longer hajj also involves travel to Mian and a return to Mecca.

Catering to pilgrims has been Mecca's main industry for over 1,300 years. In the past pilgrims would converge on the city across the desert on foot, donkey or camel, reinforcing the sense of supplication, a journey through the wilderness heightening the sense of a blessed arrival at a holy site. Today many fly into the international airport at Jeddah, travel by car and bus and then walk to the giant mosque that encircles the Kaaba. The Saudi Arabian government has spent vast sums improving access to the city and providing pilgrims with transport, food, accommodation and health facilities. It is estimated the Saudi government spent $14 billion in the past 25 years. They built the world's largest abattoir to provide meat, constructed a water factory that provides 50 million bags of fresh water and every year they transport half a million people by Saudi Air to Jeddah. The revenues from oil have allowed improved facilities and access to the city. Crowd stampedes in 1990, 1994 and 1998 also prompted changes to the spatial layout, facilitating easier movement by the vast crowds, and extra money is spent on security and surveillance to hold in check the threat of violence and terrorism by religious fundamentalists and dissident groups.

Traditionally water in Mecca came from surrounding wadis and connecting tunnels that ranged over 20 miles in the south and 60 miles in the north. Now a government, rich with oil revenues, can make water and freely give it away.

Las Vegas is also a desert city but one that hides its dry location with extravagant displays of water. The city grew up in the modern era as a remote frontier post, a stepping-off point on a journey to somewhere else. In 1905 it was a railroad town, but because of its location, further growth was difficult to imagine. The town is in the middle of the Mohave Desert, far from population centers, with precipitation less than five inches per year and summer temperatures that can soar well above 100°F. The building of the Hoover Dam, only 30 miles from Las Vegas, initially named the Boulder Dam, was a significant event in the development of Las Vegas. It began in 1929 and brought in thousands of federal workers. While there were only 5,000 jobs, almost 42,000 people came in the job-starved Depression years looking for work. The state legislature, realizing an economic opportunity, legalized gambling. The Dam, eventually completed in

1936, laid the basis for plentiful water and cheap energy. The federal government further helped the growth of the city by building a base for training pilots, later named Nellis Air Force Base, that housed 10,000 people. The federal government laid the basis of the economic development of the city but it was Mob figure Bugsy Siegel who saw its full potential. His *Flamingo Hotel* opened in 1946 and inaugurated a new gambling era. Casinos funded and operated by organized gambling made Las Vegas a destination point for more people. In 1950 the permanent population was still only 24,624. There was a reaction to the organized crime. In 1955 the state of Nevada passed legislation that required all shareholders to be licensed. It was designed to keep out known mobsters. However, it enshrined rather than replaced organized crime. In an era of corporate capitalism the large financial institutions would have to license everyone of their hundreds of thousands of shareholders. This requirement was only overturned by the 1967 corporate gaming Act that paved the way for corporate investment in the city. In the past twenty years the city has changed as megaresorts have replaced casinos. The 1989 opening of the *Mirage* marked a movement away from a model of money being made from casino gambling to more profits being generated from entertainment. While some older casinos still exist such as the *Golden Nugget*, *Four Queens* and the *Horseshoe*, other such as the *Dunes*, *Sands*, *Aladdin* and *Hacienda* were demolished and replaced with hotel and entertainment spectaculars such as the *Bellagio*, *Venetian*, *Paris* and *New York*. Even the name changes from the *Dunes* and the *Sands* to the *Venetian* and *New York* represent a change from a casual reference to the desert to a postmodern simulacrum of places from around the world. The themed hotels and casinos evoke a global fantasy world (see Figure 5.4).

Las Vegas is desert city with the highest per capita water consumption in the US. By 2005 the city had grown to over half a million and water consumption was just over 400 gallons per person. Water plays a crucial role in the fantasy presentation of the city including the spectacular presentation of hotel architecture such as the winding Grand Canal of the *Venetian*, the extravagant water displays of choreographed waters jets, the glistening water surfaces of the swimming pools and the lush green fairways of well-watered golf courses. Las Vegas defiantly negates its desert setting in a fantasy of urban spectacularization for its 35 million annual visitors, as Figure 5.5 highlights.

In Mecca, the desert is used to reinforce a sense of purity, sublimation, redemption through difficult pilgrimage, its austerity perhaps echoed in the simple and fundamental tenets of Islam and the fundamentalist interpretation of Saudi Wahhabism. Strong rules for people living in a harsh environment. In Las Vegas the negation of the dryness of the desert is a persistent theme. The waterfalls,

Figure 5.4 *Las Vegas: city of simulation*
Source: Photo by Joe Dymond

the pools and the lush greenery all signal not so much a defeat of nature but indifference. The air-conditioned rooms do not so much defy the dry heat as ignore it. The desert is significant only in its silence.

And yet there are similarities between the two cities in the desert. There is the same sense of pilgrimage, a journey to an oasis. The same promise of a life-changing experience. Located in the most inhospitable places, each city offers something to the traveler; riches and redemption, spiritual connection or easy money, enlightenment or entertainment. The desert location informs the identity and functioning of the city both in its presence and its absence.

City on the delta

The city of New Orleans owes its origins, economic rationale and a possible source of its destruction to the Mississippi. Situated close to the mouth of the

Figure 5.5 *Las Vegas fountains display the city's fantasy of extravagant disregard for water.*
Source: Photo by Joe Dymond

world's third largest river system, this location provides both the *raison d'etre* and possible annihilation for the city. The city's history, geography and future is intimately bound up with its position on the delta.

New Orleans is less a result of a careful assessment of environmental suitability and more the result of geopolitical power struggles. It began as a French city. Early French incursions into North America were along the St Lawrence and then down the Mississippi. After the French founded Quebec in 1602 they began to move along the inland riverways and waterways, lured in part by the prospect of controlling the fur trade. Traders, Jesuit priests and functionaries traveled along the Mississippi river in search of pelts, converts and allies. They sold liquor, metal tools and blankets in return for fur, religious conversion and political alliances. In 1682 Rene-Robert, Cavalier de La Salle, leader of 40 French and Indians, entered the great river from the Illinois River and moved south paddling downstream. By April they had reached the mouth of the river. On 9 April they planted a cross and raised the French standard to formally claim possession of the entire river basin.

Figure 5.6 *Similar to Sydney, Santa Monica in Los Angeles has a very strong beach culture.*
Source: Photo by Michele A. Judd

A merchant company was given title to Louisiana by the French government in 1717. The government gave the company, known as the Mississippi Company though formally entitled the Company of The West, a trade monopoly for 25 years and a directive to establish 6,000 free settlers and 3,000 slaves. This was a common practice by imperial powers to retain overall control of overseas territory but to offload the costs and everyday management of colonization to a merchant company. The city by New Orleans was founded by the merchant company in 1718. The attempt to build a trading city close to the mouth of the river encountered the watery geography of a giant delta, half marsh, half mud, a floating spongy raft of shifting vegetation. The city was located 120 miles from where the river flows into the Gulf of Mexico, at a bend in the river close to Lake Pontchartrain, a site that enabled the portage of goods from the lake to the city. It was easier to ship goods to the lake and transport them to the city than to sail up the ever-shifting Mississippi river.

The city was sited on a relatively high piece of land where French traders had already been encamped close to a Native American trail. The city's origins are thus part Native American, part French, and the joint ancestry is apparent in names. The city was named after the Duc of Orleans who was regent and ruler of France in 1718. The French wanted to call the river the St Louis, but the Native American name of Mississippi persists to this day. The French legacy is apparent in the French names along the lower Mississippi as well as the long lot land use patterns. Under the French, land holdings were divided up as long strips with narrow river frontages.

The city was an outpost of the French empire, part of a global network of colonial possessions that stretched from the Americas to Africa and Asia. French power in North America was concentrated along the St Lawrence. New Orleans was at the outermost edges, at the end of a long journey south across lakes and rivers and down the Mississippi. Other French towns along this river highway included Quebec, Montreal, Detroit, St Louis and Baton Rouge. The city was also part of wider French interests along the Gulf coast. Other cities in this Gulf cluster included Mobile and Biloxi.

Large land grants were awarded and several thousand colonists were brought from France, Canada, Germany and Switzerland. Native Americans resisted the incursions into their land, but the French military defeated them and subsequently shipped many of them off to the French colony of St Dominique as slaves. On the lower Mississippi, the colony soon evolved as a plantation economy, growing sugar, rice, indigo and tobacco. Labor was scarce so slaves were imported from Africa to work in the fields. At the center of this new colony was New Orleans.

Cartesian order – the imposition of the grid – was introduced to the North American wilderness. New Orleans was laid out as a rectangular grid of 44 blocks, eleven blocks running alongside the river and four away from the river. The city was enclosed in fortifications. At the center block immediately facing the river was the open square of Place d'Armes surrounded by government and religious buildings. This street pattern remains a distinctive feature of the French Quarter, or *Vieux Carré*, of contemporary New Orleans.

The city grew slowly at first. A 1764 map of the city shows that a third of the blocks were empty. And even by the end of the eighteenth century not all the land in the city boundaries was occupied. It took some time before reality filled in the promise of the grid. Buildings were constructed initially out of wood, to be replaced, as time and wealth accumulated, by more substantial and permanent brick buildings often stuccoed in white or yellow. Houses were built on pillars with large verandahs that wrapped around the entire floor; a design that maximized air circulation in an oppressive climate. The low water table created difficulties in construction, and the river threatened the city with regular flooding.

New Orleans and the Mississippi from three sources

As one looks upon the Mississippi, which curves and winds round New Orleans, as it does in every part of its course, and from which the title of the Crescent City is derived, we look in vain for the reasons, which prompted the choice of the site. So far as the river is concerned, the city might have been a hundred miles higher or fifty lower ... the ground is so low that the drainage runs away from the river. Immediately behind the city is a low swamp, which generates fever and disease, and which is the secret of the unhealthy condition of the place. On such low ground the city is built.

William Kingsford (1858) *Impressions of The West and South, During Six Weeks' Holiday,* Toronto: Armour, p. 54.

As far as he can see almost, up and down, the margin is lined with flat-boats, come from above, from every part of the Valley of the Mississippi. Some are laden with flour, others with corn, others with meat of various kinds, others with live stock, cattle, hogs, horses or mules. Some have traveling stores: occasionally some are to be found which are full of negroes ... along the lower part, he will see a forest of masts; higher up we may see 20 or 30 steam-boats, with their bows up against the levee, or else projecting over an "up-country" flat-boat. Every day some come from above and others depart, on excursions of one or two thousand miles, to St Louis, or Louisville, or Nashville, or Pittsburgh, or hundreds of other places. For distance is no longer thought of in this region—it is almost annihilated by steam ... and if he cast his eye down the river, he may see a whole fleet.

Baird, Robert (1834) *View of the Valley of the Mississippi,* Philadelphia: H. S. Tanner, pp. 278–281.

As sediments slide down the continental slope and the river is prevented from building a proper lobe—as the delta subsides and is not replenished—erosion eats into the coastal marshes and quantities of Louisiana steadily disappear. The net loss is over fifty square miles per year ... A mile of marsh will reduce a coastal storm-surge wave by about one inch. Where fifty miles of marsh are gone, fifty inches of additional water will inevitably surge, The Corps [of Army Engineers] has been obliged to deal with this fact by completing a ring of levees around New Orleans, thus creating New Avignon, a walled medieval city accessed by an interstate that jumps over the walls.

John McPhee (1989) *The Control of Nature,* New York: Farrar Straus Giroux, pp. 62–66.

By 1825, the fortifications were pulled down, creating opportunities for the construction of broad boulevards. Canal Street, North Rampart Street and Esplanade Avenue are sited along the wide locations of the early fortifications. As the city grew along the river lots of plantations were subdivided into grid-iron strips. The open square, broad streets and the grid became distinctive features of the city's morphology. The city grew first along the river and then north toward the lake. Initially development was restricted to the higher ground above sea level because of the fear of flooding, but improvements in pumping technology in the early twentieth century encouraged more development in the lower-lying areas of the city.

In 1763, the city and the wider province of Louisiana changed hands. In the wake of France's defeat in the Seven Years War (the French and Indian Wars as they applied to North America), the city and territory changed from French to Spanish possessions. It made little real difference. New Orleans was always at the end of a very long tentacle of French government interest and involvement. The Spaniards proved efficient administrators. They rebuilt the city after the destructive fires of 1788. The buildings of the so-called French Quarter are more Spanish design than French. But Spanish power in the region was short lived. The territory was ceded back to the French in 1801 and then two years later it was included in the Louisiana Purchase. New Orleans became an American city. It also became more strategically located. As settlers pushed west to the Mississippi and beyond, New Orleans was now at the mouth of a giant economically productive river basin. Commodities from the interior were shipped down the river through the port of New Orleans. Slaves, goods and commodities were also shipped up river.

Steam power shrunk the distance within the giant basin while economic growth increased the traffic along the giant river system. New Orleans was a pivotal hub in this new economic geography. In the early nineteenth century the city grew from around 10,000 in 1800 to 102,193 in 1840 when it became the third largest city in the nation, behind only New York and Baltimore in population size. From 1830 to 1860 it was one of the largest half-dozen cities in the entire country, the single most important export center of the whole economy with the largest slave market in the nation. The city was at its high economic point on the eve of the Civil War when technological change and new economic geographies propelled the city to a prime position. Subsequent technological changes and further transformations in economic geography undercut its strategic location. As rail replaced river traffic the city was no longer such a vital hub. As the economic center of gravity shifted ever westward and as the national economy transformed from an agricultural to a more manufacturing base, New Orleans lost its importance. Railroad cities such as Chicago grew to prominence and industrial cities such as Cleveland, Buffalo and Pittsburgh all eclipsed the crescent city. The population continued to increase, but its relative position was slipping. In 1900 the city's population had increased to 287,104 but the twentieth century saw the long slow relative decline of the city as it moved from twelfth largest city in 1900 to sixteenth in 1950, twenty-fourth in 1990 and thirty-first in 2000. New Orleans lost influence and national significance. Factories were locating elsewhere, waves of economic growth passed it by. The city remained deeply divided along lines of class and race. New Orleans seemed to miss out in the creation of a substantial middle class. The city languished in relative decline.

The river that gave the city economic prominence also threatens its destruction from two sources. The first is flooding. The history of New Orleans is one of floods, responses to floods and measures to avoid floods. When the snow melts in the upper regions of the vast river basin and rainfall is heavier than usual then the river can flood its banks. The city was first constructed only months after a major flood. And since then floods have been an integral part of the city's history. The city flooded in 1731 and 1752. In 1816 the city was flooded for a month. The flood of 1828 prompted the mandating of taxes to pay for levees. In 1849 the city was flooded for 48 days. The floods of 1850 prompted the federal government to pay for levees. The river had major floods in 1882, 1884, 1890, 1891, 1898, 1903, 1912, 1913, 1922 and 1927. When 20 inches fell on the city in 1995 the heavy rain caused the city to flood.

One response was to build levees to keep the water in check. Since the first four-foot levees were built in 1722, levee construction has been the predominant response to the threat of flooding. Keep the river in high channels was the solution. One influential theory in the nineteenth century argued that levees would not only keep floods at bay but by channeling the water in a narrow area would increase flow and hence scour out the riverbed bottom. The bankruptcy of the policy was revealed by the great flood of 1927 when more than one million people were flooded from their homes. Thereafter spillways, that would release the floodwaters in a controlled way, supplemented levees.

Urbanization created a greater risk of flooding. As the amount of impermeable surface increased in both the city and throughout the entire Mississippi river basin, the risk of flooding also increased. Every new parking lot and residential development reduces the gradual absorption properties of the more permeable surfaces and increases the likelihood of flooding.

Because the river keeps depositing sediment, river levels rise and so, to be effective, levees have to be raised. In order to be safer, levees have to be increased in size, which means more people are thus located below the river in the event of flooding. The River Mississippi now flows through New Orleans as an elevated water highway. Rain in the city has to be pumped out and this led to a steady sinking of the city below sea level. The more the city sank the more pumping was necessary and the higher the levees needed to be. Residential development followed the space-packing process common to other private market societies. The rich get the best sites with lower-income groups getting less desirable places to live. In New Orleans the most powerful groups had commandeered most of the high ground. There was a connection between altitude and class. The poorest in the city live in neighborhoods like the Ninth Ward that are well below sea level and most vulnerable to floods. The higher elevation Central Business District and the more upmarket Garden District and French Quarter escaped the worst flooding. The lower-lying neighborhoods of poor, black people in rented accommodation were the most severely affected by the post-Katrina flooding. There are exceptions. Low-lying Lakeview was a predominately white middle- and upper-income area that also flooded in 2005.

Levees need continual maintenance and monitoring. Yet levee maintenance lacks political support until it fails. It was levee failure that led to the catastrophic flooding after Hurricane Katrina in 2005 that killed over a thousand people and devastated the city.

The second problem for New Orleans is the dynamic quality of the Mississippi River. It is best likened to a giant hose. As water squirts out, the hose moves from side to side. Over the long term the river has shifted its position as it deposits sediment and creates new river courses. The river is continually changing course because of the changing topography and the gravitation pull of searching for the quickest way to the sea. Left to themselves the giant loops of a meandering river would constantly fantail across a very broad area. The channelization of the river keeps things in place but at a cost. Three hundred miles upstream from the mouth of the river, the US Army Corps of Engineers had to build giant dams to keep the water flowing down the Mississippi rather than into the Atchafalaya, which has a sharper gradient and quicker route to the sea. Without the humanly engineered system the river would divert well to the west and south and New Orleans would no longer be a river city. The Old River Control Project was begun in 1954 to keep the water split between the Mississippi and the Atchafalaya at 70/30. The project – conservative estimates put the final cost at a billion dollars – involved the construction of earthen dams and intricate flood control systems. The control of nature is achieved at tremendous cost and constant vigilance.

Guide to Further Reading

The city–site nexus is a major theme of the new urban environmental history that is emerging. For a review from one country see the extensive literature review in Tarr, J. A. *Urban History and Environmental History in the United States; Complementary and Overlapping Fields.* http://www.h-net/org/~environ/historiography/usurban.htm (accessed 25 July, 2006).

For more on Sydney

Connell, J. (ed.) (2000) *Sydney; The Emergence of a World City*, South Melbourne: Oxford University Press.

Hall, R. (2000) *Sydney: An Oxford Anthology*, South Melbourne: Oxford University Press.

Morris, J. (1992) *Sydney*, London: Penguin.

Spearritt, P. (2000) *Sydney's Century*, Sydney: UNSW Press.

For more on Las Vegas

Ferrari, M. and Ives, S. (2005) *Las Vegas: An Unconventional History*, New York: Bullfinch.

Rothman, H. (2003) *Neon Metropolis: How Las Vegas Started the Twenty-first Century*, London and New York: Routledge.

Schumacher, G. (2004) *Sun, Sin and Suburbia: An Essential History of Modern Las Vegas,* Las Vegas: Stephen Press.

Mecca as pilgrimage site

Wolfe, M. (ed.) (1999) *One Thousand Roads to Mecca: Ten Centuries of Travelers Writing About The Muslim Pilgrimage*, New York: Grove.

On New Orleans

Colten, C. (ed.) (2000) *Transforming New Orleans and Its Environs*, Pittsburgh: University of Pittsburgh Press.

Colten, C. (2004) *An Unnatural Metropolis; Wresting New Orleans from Nature*, Baton Rouge: LSU Press.

Lewis, P. F. (2003, 2nd edn) *New Orleans: The Making of an Urban Landscape*, Santa Fe: Center for American Places.

6 Cities, environmental hazards and disasters

Hurricane Katrina: natural disaster?

On Thursday August 25, 2005, a tropical storm reached the coast of Florida. As winds reached over 73 mph it was upgraded to hurricane status. Hurricane Katrina, as it was now officially designated, quickly passed over the southern-most tip of Florida, then moved into the Gulf of Mexico where it strengthened as it soaked up the warm gulf waters. At 11 am on Sunday August 28 wind speeds exceeded 170 mph. Katrina was now a fully-fledged category 5 hurricane, capable of inflicting ferocious damage and enormous harm.

Hurricanes are not unusual events. Geological evidence from lakebed sediments suggest that they have been a regular occurrence for at least 3,000 years and probably much longer. They are an integral part of the environment of the Caribbean, appearing in ancient Mayan hieroglyphics. The ancient Mayans built most of their settlements away from the coast; they knew about hurricanes and organized their settlements accordingly. Cities were built away from the dangers of the strandline. Nature was accommodated in human affairs, a due recognition of the power and intensity of storms and natural calamities. In contrast, today we are building more of our settlements along the coast. Coastal properties and beach locations are prime sites, attracting development and growth as popula-tions and investments congregate close to the shoreline. Rather than conforming to natural systems we take a more arrogant view than the Mayans. Nature is to be subdued, controlled, managed. And in large measure we have been extraordi-narily successful. We have transformed mangrove stands into beach resorts, swamps into cities and strandlines into suburbia. Coastal development in hurri-cane zones is just one more manifestation of a deeper and wider reorientation of society and nature. We do not accommodate nature but taunt it by building along earthquake fault lines, placing developments on top of unstable mountains, and constructing cities in the middle of hurricane zones. We ignore our increased knowledge of predictable environmental hazards, accepting short-term benefits and discounting longer-term possible costs. This is part of a deeper and wider sense that the environment is not something that we accommodate to, but something we subdue. In part it is a rational position in the rich West. In the past two hundred years an environmental revolution has been wrought through science and technology. We have overcome many of the constraints imposed by the environment. Electricity turns night into day, air conditioning transforms humid hot air into cool dry air, heating warms cold air. We have the power to transcend many of the limits imposed by the world we live in. We inhabit a human modified environment. And as more people live in cities and get their food from the grocery store and their weather information from the television, the sense of the environment as a limiting factor gradually disappears. More of people's time and environmental experience, especially in the rich world, is in the socially constructed environment of the home; the office and the shopping mall where climate is controlled and temperatures are fixed. From this perspective, weather patterns become an external event seen through the car window or on the television, and even hurricanes become an occasional inconvenience, not a determinant of settlement patterns. Where the Mayans listened to the voice of the environment we ignore it.

The path of Hurricane Katrina was accurately predicted. As early as Friday 26 August people knew it would make landfall close to the city of New Orleans in southern Louisiana. Hurricanes are not unusual events in New Orleans. Since 1887 at least 34 hurricanes and 25 tropical storms have passed within 100 miles of the city. Since 1960, 17 hurricanes have passed within a hundred miles. New Orleans had been spared destruction. By the time Hurricane Isidore reached close to New Orleans in September 2003, wind speed had fallen to 63 mph and it had been downgraded to tropical storm status. Hurricane George in September 1998 reached wind speeds of 110 mph as it passed over Puerto Rico, but by the time it had crossed the Gulf and moved towards New Orleans wind speed had declined to 57 mph. For the people of New Orleans hurricanes were no big deal. And here we encounter a paradox. Major hurricanes are extreme events that will inevitably take place sometime. The longer that they do not occur, and the more complacent people become, the more likely it is that they will occur. Imagine an event that will take place, on average, once every hundred years. You know it will occur within this time but you do not know when. As time passes and the event does not happen the more you are likely to forget about it or ignore it. But the passage of time that creates complacency that it will not occur also makes the event more likely to occur. We believe tomorrow is going to be much like today and yesterday.

Hurricanes cause three main forms of damage; rain, wind and flood. Hurricanes, especially the large slow-moving ones, can produce downpours of up to 25 inches in 30 hours. The high winds tear buildings and other structures away from their moorings. Everything not securely tied down (and even those that are) becomes subject to the vagaries of the wind and even secure structures experience wind damage. The high winds also generate waves. Storm surges associated with the high winds can reach over 20 feet above normal sea level, causing massive flooding. For most coastal cities flooding is a serious problem, for a city below sea level it is potentially catastrophic. Much of New Orleans is below sea level.

Hurricane Katrina made landfall just east of the city at 6:10 am on Monday August 29. Winds had dipped in strength to 145 mph. It was not the ferocious winds that damaged the city but the storm surge that breached levees in the city. The city was flooded when parts of levees at 17th Street and Industrial Canal collapsed. Almost 80 percent of the city was flooded, in some cases by water over 20 feet in depth. An estimated 1,000 people were killed, most of them drowned by the rapidly rising floodwaters. Much of the city was destroyed in the flooding that followed the hurricane.

When is a disaster a natural disaster? At first blush Katrina seemed like a natural disaster. A hurricane is a force of nature. But it was a force of nature whose impacts and effects were mediated through the prism of socio-economic power structures and arrangements. The flooding of the city was caused by the poorly designed levees that could not withstand a predictable storm surge. It was not Katrina that caused the flooding but shoddy engineering, poor design and inadequate funding of vital public works. Storms surges are neither unknown nor unpredictable in New Orleans. Yet the levees were poorly constructed with pilings set in unstable soils. They should have been built to withstand 15 foot surges but had settled in many places to only 12 to 13 feet above sea level.

The storm surge itself was particularly severe because of the loss of wetland. For the previous 20 years wetland was lost in the Gulf Coast at a rate of 24 square miles per year. The wetlands have a deadening effect on storms as they absorb much of the energy and water the storm brings. It is estimated that wetlands reduce a storm surge by one foot for every 3 miles of marsh. The loss of wetlands around New Orleans, something predictable and knowable, added to the potency of the storm surge. The engineered landscape in and around the city exacerbated the storm surge, channeling more water into the city. The Mississippi River Gulf Outlet (MRGO) was built in 1965 to shorten ship travel from the city to the Gulf. Although rarely used it has an enduring ecological affect. It allowed the penetration of salt water into wetlands, which destroyed 25,000 acres of marsh and funneled the storm surge directly into the city. Like the disappearance of the wetlands, the environmental impact of the Gulf Outlet was predictable and known.

The mayor of the city issued a voluntary evacuation on Saturday August 27 and a mandatory evacuation the next day. Those with cars were able to get out but there was little provision for the most vulnerable; those with no access to private transport were abandoned. While the more affluent could leave, the very poorest, the most disabled, the elderly and infirm were trapped. Between 50,000 and 100,000 were left in the city when the hurricane struck and the levees failed. Some made their way to the Superdome and the Convention Center, which by Wednesday 31 August were housing between 30,000 and 50,000 people. They remained for days, a stunning indictment of social and racial inequality in its starkest and bleakest form.

The effects of Hurricane Katrina on the city were socially and racially determined. Flooding disproportionately affected the poorest neighborhoods of the city. The more affluent, predominantly white sections of the city, such as the French Quarter and the Garden District, were at a higher elevation and escaped flood damage. The flooded areas were 80 percent nonwhite. The hardest hit neighborhoods were nonwhite and most of the high poverty tracts were flooded. The racial and income disparities in the city were cruelly reflected in the pattern of flood damage. In that sense the "natural" disaster appears in closer detail as a social disaster. As a Congressional bipartisan report noted, "Katrina was a national failure, an abdication of the most solemn obligation to provide for the common welfare."[128] Figures 6.1 and 6.2 show the devastation of Katrina and its lingering impacts more than 18 months after the hurricane.

Scales of disaster

Two of the main disaster types, hurricanes and earthquakes, have scales that allow measurement and comparison.

The Saffir-Simpson hurricane is a 5-fold categorization based on wind speed. The scale also gives indications of possible storm damage.

Category 1: wind speed of up to 95 mph, storm surge above 4–5 feet; little damage to secured buildings.
Category 2: wind speeds up to 110 mph, storm surge up to 8 feet above normal; damage to vegetation, some flooding.

Continued

Category 3: winds up to 130 mph, storm surge up to 12 feet above normal;
flooding possible up to 15 miles inland, damage to property, evacuation of
low-lying residences may be required.

Category 4: winds up to 155 mph, flood surge up to 18 feet above normal;
extensive damage to property, massive evacuation of residential areas up to
6 miles from coast may be necessary.

Category 5: winds greater than 155 mph, storm surge more than 18 feet above
normal; massive evacuation up to 10 miles from shoreline may be required.

The scale is not one of equal steps of arithmetic progression. Storm surge destruction is closely related to the square of the wind speed; greater wind speed increases the possible damage by the square of the difference, not the arithmetic difference. A category 5 hurricane can inflict 16 times more damage than a category 3 hurricane.

Charles F. Richter devised the scale that bears his name in 1934. It measures the amplitude of the largest seismic wave. Values are recorded from 1 to 10 in a geometric progression. An earthquake of magnitude 1 has the destructive force of 6 ounces of TNT. A magnitude 8 earthquake has the destructive force of 6 million tons of TNT.

Richter Magnitude	Effects
2.5 or less	Usually not felt
2.5 to 5.4	Sometimes felt, minor damage
5.5 to 6.0	Slight damage
6.1 to 6.9	Damage in more densely populated areas.
7.0 to 7.9	Serious damage.
8.0 or greater	Major earthquake with very destructive powers

Disasters and cities

As the title of one edited book about Katrina proclaims, *There is No Such Thing as a Natural Disaster*.[129] It seems more appropriate to use the term environmental disaster rather than natural disaster as there is nothing "natural" about the effects of a disaster, rather they are first and foremost social. Disasters reveal our social fault lines and make our political structures clearly visible. There is a

Figure 6.1 *Waiting for clean up in New Orleans. More than 1½ years after Hurricane Katrina, some neighborhoods have yet to be cleared of debris. This has delayed rebuilding efforts.*
Source: Photo by Stefanie and Katherine Garry

growing literature on the relationship between cities, hazards and disasters that raises issues of vulnerable cities, at risk urban dwellers, the social differences in how urban disasters are experienced and the resiliency of cities to bounce back from a disaster.[130]

Disaster results from vulnerability to environmental hazards. Hazards include earthquake, floods, windstorm, fire, and landslides, flooding and many others. There is a geography to this vulnerability. Some cities are more vulnerable to certain hazards than others because of their location; Mexico City is more vulnerable to volcanic eruption than New York City. Table 6.1 notes the ten largest cities and their associated hazards. But the vulnerability to hazards is more socially constructed than geographically determined. Poorly planned urban growth, deforestation and poor medical provision are just some of the factors that increase the susceptibility of hazards becoming disasters. Many cities and particularly the poorer citizens are vulnerable to environmental hazards that can

Figure 6.2 *Waiting to rebuild in New Orleans. Rising frustration is expressed by signs of protest and requests for assistance in rebuilding.*
Source: Photo by Jessie Goldman

Table 6.1 Cities and hazards

City	Population (2005)	Hazards
Tokyo	35.2 million	earthquake, storms, tornado, storm surges
Mexico City	19.4 million	earthquake, volcano, storms
New York	18.7 million	earthquake, storms, storm surge
São Paulo	18.3 million	storms, flood
Mumbai	18.2 million	earthquake, storms, flood, storm surge
Delhi	15.0 million	earthquake, storms, flood
Shanghai	14.5 million	earthquake, storms, flood
Calcutta	14.3 million	earthquake, storms, tornado, flood, storm surge
Jakarta	13.2 million	earthquake, flood
Buenos Aires	12.6 million	storms, flood

Source: Adapted from Chafe, Z. (2007) "Reducing natural disaster risk in cities", *State of The World: Our Urban Future*, New York: Norton, pp.112–129.

turn into disasters. The risks are often greatest in cities because there is a greater concentration of both people and hazards, an increasing number of hazards and the possibility of hazards interacting with each other – as in floods causing disease outbreaks or earthquakes causing fires. Disasters are especially prevalent in cities where emergency and response mechanisms are often too little too late.

In the 1980s, 177 million people were affected by disasters; by 2002 this had increased due to population growth, rapid urbanization, environmental degradation and climate change to almost 270 million. Of those affected, 98 percent live in low-income countries. In 2005, 430 environmental disasters killed almost 90,000, the vast majority in low-income countries. Disasters differentially impact the poorest in the cities of the developing world and they play a part in sustaining poverty and inequality.[131] Analyzing the connections between hazards and disasters and outcomes in cities makes an important link in nature–society relations.

Figure 6.3 Hurricane Ivan, 2004. After wrecking havoc in the Caribbean, especially on the island of Grenada, Hurricane Ivan moves toward the mainland USA.
Source: http//images.google.com/imgres?imgurl=http://www.nnvl. noaa.gov/hurseas2004/ivan1945zB-040907-1kg12.jpg&imgrefurl=http: www.nnvl.noaa.gov/cgi-bin/index.cgi%3Fpage%3Dproducts%26category% 3DYear%25202004%2520Storm%2520Events%26event%3DHurricane% 2520Ivan&h=1199&(public domain)

Urban environmental disasters show, in the starkest terms, the vulnerability of cities. In the past decade or so, more attention has focused on the vulnerability of cities to environmental disasters. There is a rereading of history to excavate the role of disasters in the evolution of cities, as well as an examination of current levels of risk, preparedness and responses. In this more anxious age there has been re-examination of cities as sites of disasters. In the wake of 9/11, tsunamis, Hurricane Katrina and terrorist attacks in cities around the world, there is a more marked sense of the vulnerability of life and the precarious nature of urban civilization. This perspective has shaped urban studies focused on urban disasters.

For example, in his book *Ecology of Fear* Mike Davis examines some of the disasters that continually threaten the Los Angeles metropolitan region. From the storms that sweep across the LA basin from the Pacific, to the wildfires of summer and the landslides that push expensive houses into the valley or sea below, Davis presents a picture of a city in an ecological disaster zone. Although exaggerated, Davis does point to the often fragile presence cities have on the surface of the earth.[132]

Mark Pelling looks at cities in the Caribbean as sites of disasters. With reference to Bridgetown (Barbados), Georgetown (Guyana) and Santo Domingo, (Dominican Republic), he highlights the issue of social vulnerability but also examines the resiliency of cities to rebuild after a disaster strikes.[133] We can identify a number of possible urban environmental disasters including fire, flood, earthquake and war. Table 6.2 provides a very small fraction of the recent major disasters that have affected cities all over the world. In the remainder of this chapter we will consider examples from these disaster categories.

Table 6.2 *Recent major urban disasters*

City	Major Disaster	Date	Reported deaths
Bam	earthquake	2003	26,300
Dhaka	flood	1970	1,400
Paris	heatwave	2003	14,800
Istanbul	earthquake	1999	15,000
Kobe	earthquake	1995	6,400
Tangshan	earthquake	1976	242,000

Old books, recurring themes

The recent focus on urban environmental disasters is only the latest manifestation of a long fascination with disaster. In 1913, areas in the Midwest of the US were hit by a series of tornadoes, floods, storms and blizzards. To match the near apocalyptic events Logan Marshall wrote a marvelously overwritten book, *Our National Calamity of Fire Flood and Tornado*. The elaborate style makes it a product of its time but raises general issues that still resonate down the years.

> Man is still a plaything of Nature. He boasts loudly of conquering it; the earth gives a little shiver and his cities collapse like a house of cards ... He imprisons the waters behind a dam and fetters the current of the rivers with bridges; they bestir themselves and the fetters snap, his towns are washed away and thousands of dead bodies float down the angry torrents ... He burrows into the skin of the earth for treasure, and a thousand men find a living grave.

Source: Logan, M. (1913) *The True Story of Our National Calamity of Flood Fire and Tornado*, Lima Oh: Webb Book and Bible, p.11.

Fire

The Great Fire of London began on Sunday morning, September 2 1666. The fire started in Pudding Lane and swept through the city for five days until the wind died down and the use of gunpowder explosions created firebreaks. The fire was not the first. Previous fires in 1133 and 1212 also caused extensive damage. The city, full of timbered high-density buildings and the extensive use of candles and wood and coal fires for heat and light, was a conflagration waiting to happen.

The Great Fire caused major damage; 13,200 houses were destroyed, as were 87 churches and many fine public buildings. Over 200, 000 people were displaced. The official death toll was only 6 but many more are presumed to have died especially from smoke inhalation. The fire swept across 436 acres burning almost all the buildings inside the medieval walls.[134]

The Great Fire also led to the planning and rebuilding of London. In the fire's aftermath new building codes were introduced. Wood buildings were replaced with brick and stone buildings. Laws restricted dwellings to two floors in lanes and three in the larger streets. New laws also required owners to obtain insurance and the new insurance companies quickly realized that employing men to put out fires could minimize their losses. Private fire-fighting companies were established, the forerunner of public fire protection. The architect Christopher Wren planned much of the reconstruction of London and built 51 new churches including his masterpiece, St Paul's Cathedral. From the ashes of the Great Fire a new London was constructed, a more modern London in terms of building design, street layout and in the provision of collective services such as insurance and fire protection.

Throughout history fires have been an urban hazard. Cities composed of wooden structures with widespread use of flame power sources are an ideal formula for fire. Small-scale fires were regular occurrences. Epic fires were rarer. A small fire in Chicago in fall of 1871 quickly turned into a widespread conflagration. It began around 9 pm on Sunday, October 8, 1871 and lasted until the morning of October 10 when a light rain helped to douse the remaining flames. The fire killed 300 people, left 100,000 homeless and destroyed most of the downtown. Yet the fire, although a traumatic event, barely seemed to halt the growth of the city. Within five years the city was rebuilt and in 1891 hosted the Columbian Exhibition that showcased the City Beautiful Movement and celebrated the rebuilt city.[135]

Eyewitness account of 1871 Chicago fire

Horace White, editor at the *Chicago Tribune* was an eyewitness to the Chicago fire of 1871. He wrote the following shortly after the fire on 14 October 1871:

There was a great light to the southwest of my residence, but no greater than I had frequently seen in that quarter, where vast piles of pine lumber have been stored all the time I have lived in Chicago, some eighteen years. But it was not pine lumber that was burning this time. It was a row of wooden tenements in the South Division of the city, in which a few days ago were standing whole rows of the most costly buildings which it has entered into the hearts of architects to conceive. I watched the increasing light for a few moments. Red tongues of light began to shoot upward; my family were all aroused by this time, and I dressed myself for the purpose of going to the "Tribune" office to write something about the catastrophe. Once out upon the street, the magnitude of the fire was suddenly disclosed to me.

The dogs of hell were upon the housetops of La Salle and Wells streets, just south of Adams, bounding from one to another. The fire was moving northward like ocean surf on a sand beach. It had already traveled an eighth of a mile and was

Continued

far beyond control. A column of flame would shoot up from a burning building, catch the force of the wind, and strike the next one, which in turn would perform the same direful office for its neighbor. It was simply indescribable in its terrible grandeur ... I went to the "Tribune" office, ascended to the editorial rooms, took the only inflammable thing there, a kerosene lamp, and carried it to the basement, where I emptied the oil into the sewer. This was scarcely done when I perceived the flames breaking out of the roof of the court house, the old nucleus of which, in the center of the edifice, was not constructed of fireproof material, as the new wings had been. As the flames had leapt a vacant space of nearly two hundred feet to get at this roof, it was evident that most of the business portion of the city must go down, but I did not reflect that the city water works, with their four great pumping engines, were in a straight line with the fire and wind. Nor did I know then that this priceless machinery was covered by a wooden roof. The flames were driving thither with demon precision.

Billows of fire were rolling over the business palaces of the city and swallowing up their contents. Walls were falling so fast that the quaking of the ground under our feet was scarcely noticed, so continuous was the reverberation. Sober men and women were hurrying through the streets from the burning quarter, some with bundles of clothes on their shoulders, others dragging trunks along the sidewalks by means of strings and ropes fastened to the handles, children trudging by their sides or borne in their arms. Now and then a sick man or woman would be observed, half concealed in a mattress doubled up and borne by two men. Droves of horses were in the streets, moving by some sort of guidance to a place of safety. Vehicles of all descriptions were hurrying to and fro, some laden with trunks and bundles, others seeking similar loads and immediately finding them, the drivers making more money in one hour than they were used to see in a week or a month. Everybody in this quarter was hurrying toward the lake shore. All the streets crossing that part of Michigan Avenue, which fronts on the lake (on which my own residence stood) were crowded with fugitives, hastening toward the blessed water.[135]

Source: Original newspaper article accessed at: http://www.nationalcenter.org/ChicagoFire.html

Once a fire has been extinguished, then, if the city is large enough and its economy is buoyant enough, the city can be rebuilt. The Great Fire destroyed medieval London but the rebuilding laid the basis for modern London.

Chicago survived the great fire of 1871 to come back bigger and more confident than ever. Another response is the institutional rearrangements that mitigate a repeat of the fire hazard. Insurance and fire companies are created, building codes are enforced to both plan for and negate repetition of the tragedy.

In some cases fires can initiate major legislation and new policy. At the beginning of the twentieth century, New York City experienced rapid industrial growth. Around the garment district many factories employed female immigrant labor in poor working conditions. A fire that started on March 25, 1911, in the Triangle Shirtwaist factory on the eighth floor of a building killed 146 people. It was particularly horrific as the mostly young women were trapped behind locked doors; their plight visible and audible from the streets below. The next day more than 100,000 mourners passed through the temporary morgue that was set up. The official response to the widespread public revulsion included the indictment of the factory owners and the establishment of a Factory Investigating Commission that over the next three years enacted 36 new safety laws for the city. In the wake of public reaction to the fire came new forms of regulation and safety concerns that were soon imitated by other cities throughout the country. The reaction to spectacular disasters is often a pivotal point in urban public policy.

Disasters can also influence public opinion and public behavior. In 1987, a discarded match ignited litter and grease underneath a wooden escalator at Kings Cross in London's Underground. The fire quickly spread and 31 people were killed. In response, there were the policy changes – all wooden escalators were replaced and automatic sprinklers and heat detectors were installed in escalators. But there was also a major change in public attitudes. A no smoking policy, introduced two years earlier but seldom acted upon, was more strictly enforced after the fire. And in stations as well as in trains smoking became a socially unacceptable thing to do in the Underground.

The risk of fires is minimized as stone, brick and concrete replace wood in buildings, and as cleaner fuel sources provide more light and power. But there is always the danger of fires from industrial sources. Chemical fires are a hazard in certain industrial areas. On June 1 2006, an explosion in a chemical plant in Teeside in northern England, heard up to 20 miles away, caused a fire involving a deadly mixture of hydrogen, nitrogen and ammonia. Two people were injured, roads were sealed and local residents were warned to keep windows and doors closed. The risk of such chemical fires is always present. The effects are mitigated by land use planning that segregates out hazardous facilities well away from centers of population.

Another source of fires is caused by urban/suburban growth into rural and wilderness areas which are susceptible to fire. Bushfires, brush fires and forest fires have long been a feature of ecosystems. Fires caused by lightning are an environmental fact of many arid, semi-arid and dry ecosystems. These fires are natural in that they may occur naturally but they are social in that their effects are exacerbated by urban growth which has pushed settlement into fire hazard areas. Cities in places such as California and much of Australia are so affected.

The Oakland Hills of Oakland, California, were badly damaged by a fire in 1991. Several wooden-frame houses sat in an area of dry chaparral. A fire started from rubbish quickly spread in the tinder-dry environment. The resulting fire killed 25 people, and destroyed 2,843 houses. Much of coastal California is semi-arid and susceptible to fire, and these become urban hazards as suburban growth snakes its way into the less urbanized, more vegetated and hence more at risk environments. At the end of summer with dry vegetation and high Santa Ana winds, southern California is particularly vulnerable. In late October 2003 fires scorched parts of San Diego County in California as two wildfires killed 16 people and destroyed 2,427 structures.[137]

Australia is a dry continent and the fire damage between 1969 and 1999 was estimated at an annual cost of $A77 million. Most of Australia's most devastating bushfires occur in eucalyptus forests in the suburban fringes of major cities. In the hot arid summers eucalyptus trees produce a flammable gas that is easily ignited and quickly spreads. A bushfire is in fact a series of minor explosions as the gas catches fire. The trees are adapted to fire: their bark and leaves can withstand high temperature and after a bush fire the vegetation will return. However, as suburbanization has shifted more people out towards the bush, these fires have more devastating effects on humans and their property. As recently as 1983, 76 people died in bushfires in South Australia and Victoria. In some cases arsonists have intentionally lit fires. But "natural" bushfires continue to occur. In January 2003, fires caused by lightning strikes struck the national capital of Canberra. The dry vegetation and high winds provided the perfect conditions for a major fire that burned for almost a week in the surrounding bush before entering the suburbs on January 18. Four people were killed and 816 houses destroyed. An official inquiry suggested a greater use of controlled burning to minimize the build up of dry vegetation and reduce the possibility of widespread bush fires. Figure 6.4 depicts a common sight in Australia, a sign indicating the potential for forest fires.

As suburbs spread into drier environments the risk of fire increases. In the longer term controlled burning may provide prevention. But perhaps there is a need to

Figure 6.4 *Australian Bush Fire Alert. Bush fires are a major urban hazard in Australia.*
Source: Photo by John Rennie Short

reconsider the siting of residential housing in fire risk areas; this poses a more problematic political conundrum with long-term fire safety considerations often outweighed by development interests and the forces of housing development and urban growth.

Wildfires in Los Angeles

Why is the American West burning again, and what might we do about it? There is a short answer to the first part: The American West has large wild-land fires because its extensive wild lands are prone to burning. Planning policy is much harder and requires us to consider fire history.

Natural fire regimes beat to the rhythm of cyclic wetting and drying: it must first be wet enough to grow combustibles and then dry enough to get them ready to burn. Wet forests therefore normally burn during dry spells, deserts after rains. Fire also demands a spark, and under wholly natural circumstances, this means dry lightning. The eastern United States has wet lightning, which normally accompanies dousing rain; only in Florida do thunderstorm days and lightning-kindled fires routinely overlap. The West has dry lightning – and that is why, with or without people, significant fractions of the American West will burn.

Continued

The fire "problem" resides, apparently, in the West. Why? The obvious reason is that the place is intrinsically fire-prone. The deeper reason comes from the second force: it is that the American West experienced what a historian might call an "imperial" narrative. In the 19th century, state-sponsored conservation policies encountered a landscape that had become largely emptied because the indigenous peoples had been driven off by disease-driven demographic collapses, wars, and forced relocations. It thus became possible, during that historical vacuum, for the young federal government to establish "public" lands that would exclude agricultural settlement. In doing so, it created a habitat for free-burning fire. [...]

The problem that has grabbed public and political attention is the spectacle of burning houses – the problem the agencies call the "wild land-urban interface fire." These fires might better be called "intermix fires." They occur in lands whose use has become scrambled into an ecological omelet, involving abandoned agricultural land as well as public preserves. Their existence and the hazards they pose are simply the result of unmanaged growth: the untrammeled growth of natural vegetation and the uncontained growth of our increasingly far-flung suburbs. The wild and the urban have become the matter and antimatter of the American landscape. When they collide, we should not be surprised by the occasional explosion.

Source: Pyne, S.J. (2001) "The fires this time and next", *Science* 294 (5544): 1005.

Floods

Many cities are close to water. Flooding has long been a hazard of such locations whether it be inundations by the sea or river floods. The historian John Barry tells the tale of the great Mississippi flood of 1927 which inundated 27,000 square miles and flooded the homes of nearly one million people. The flood was in part caused by steady rains through the entire river basin for months beforehand. The river flooded for 1,000 miles from Cairo, Illinois to New Orleans, Louisiana.[138]

The Mississippi has always flooded. What made the river banks so fertile was the steady deposition of alluvium by the river's changing course as it meandered toward the sea. The flooding became an "urban" hazard as more people settled and moved to the delta regions, attracted in part by the rich soils. Attempts to control the river led to the constructing of levees that effectively channeled the river, which in turn made the floods that much more devastating. Barry reveals that the response to the flood had important social and political effects. The immediate response reflected existing social and economic power. The evacuation of many black people in the delta region of Mississippi was delayed because the white plantation owners believed that black sharecroppers, often heavily indebted, would not return. A cabal of New Orleans business leaders purposely dynamited the levees of two parishes downriver of the city to protect their interests. Responses to the flood included a change in river management away from the levees-only policy and a greater use of spillways. It also brought about profound socio-economic changes including a recognition that the federal government should play a larger role. And it ushered in a change in power relations. In the immediate aftermath of the 1927 flood, the great migration of blacks from the delta to the cities of the north began and in Louisiana the populist power of Huey Long overturned the old patrician ways.

Flooding is a regular part of many river ecosystems, but can be aggravated by land use changes associated with rapid urbanization. Seoul, South Korea, has endemic flooding problems caused by river overflows from the river Han and its various tributaries, in the wake of heavy rains and rapid snow melt. Between 1960 and 1991 131 people died in floods. An analysis of the floods by Kwi-Gon Kim shows that the loss of green space leads to more flooding. Seoul's rapid growth, from a population of 2.4 million in 1960 to 10.9 million in 1990, meant a significant loss of land in agriculture and forestry and an increase in urban land uses. This shift from permeable to more impermeable surfaces raised the risk of flooding. Kim's analysis shows that the presence of green space explained 31 percent of the variation in deaths and building damage between the different wards of the city.[139]

In the cities of the rich, world flood hazards, while not eradicated, have been minimized by extensive and expensive flood control measures as well as organized rescue and recovery systems. However, in the poorer cities of the world this is less of an option. The capital city Dhaka, in Bangladesh, is located on a flood plain, with much of the city only several meters above sea level. In 1998, almost a million people were flooded from their homes. The city is located on the north bank of Buriganga River in the Ganges delta. It is has grown enormously from approximately 335,926 people in 1951 to 12.5 million in 2001. Bangladesh is one of the poorest countries in the world with a gross national income per capita in 2004 of only $440. The equivalent for Switzerland, US, UK and Netherlands are, respectively, $48,230, $41,400, $33,940 and

$31,700. Flooding is a regular occurrence in the city, especially during the monsoon season from July to September, and has become more marked as low-lying areas of the city have experienced more residential and commercial development. In addition, many of the city's poor have constructed shantytowns on steep slopes and hillsides, or in the low-lying parts of the city. Drainage backups and storms regularly lead to inundations of the low-lying parts of the city. There are also cases of more dramatic flooding. When the monsoon rains are particularly intense the rainfall overwhelms the drainage system. Mudslides can also occur, wiping away precariously perched shantytown dwellings. In 1988, a flood covered 78 percent of the city in water. In September 2004 more than half of the city was flooded including all the main roads. International aid donors have funded a Dhaka City Flood Control and Drainage Project that relies heavily on western engineering solutions, including the construction of expensive embankment and pumping stations. The operational costs are a heavy burden for such a poor country. The continual flooding of Dhaka is thus a function of unplanned urban growth into low-lying vulnerable areas, highly populated sub-standard housing which make the poor vulnerable to flood, mudslides and fires, and the poverty of the country that makes it difficult to afford efficient flood-prevention systems common to richer countries such as the Netherlands.

The effects of monsoon flooding are also impacting the ability of Bombay/Mumbai to position itself as global city. The city's 150-year old drainage system was overwhelmed by severe monsoon flooding in 2005 and 2006. In 2005 floods killed over 1,000 people. In 2006 the exceptionally high monsoon rainfalls incapacitated the city's transportation infrastructure, delaying flights and paralyzing road and train travel. The inability to deal with monsoon flooding is dampening the 18-million-population city's ambitious plan to become a hub in the financial network of global commercial centers.

Earthquakes

On November 1, 1755 the city of Lisbon in Portugal was hit by a massive earthquake estimated at between 8.5 and 9.0 on the Richter scale. The epicenter was located in the Atlantic Ocean and the city was swept by three tsunamis caused by the underwater tectonic movement. Soon fires also engulfed the city. This major disaster destroyed most of the city and killed almost 90,000 people in a city that had a population of 275,000. The city was rebuilt but never fully reclaimed its former pivotal position in the global urban network. The decline of Portugal as a major imperial force was also part of the story, but the Lisbon

Figure 6.5 Kobe Earthquake, January 1995. A completely destroyed concrete-frame office building. Such collapses blocked streets and made it difficult for firefighters to extinguish flames. Three hours after the earthquake there were at least 150 significant, simultaneous fires in the city. The only water available at this time came from tanker trucks. Although there was significant structural damage to building in Kobe, Japan, few lives were lost and rebuilding was swift and efficient.
Source: Photo credit: Dr Roger Hutchison
http://www.ngdc.noaa.gov/seg/hazard/img/200_res/21/21_431.tif

earthquake stands as testimony to the destructive power of earthquakes and the possibility of consequent urban decline.

San Francisco tells another story, one of resilience. On the morning of April 18, 1906, the city was shaken by an earthquake that registered 8.25 on the Richter scale and destroyed 25,000 buildings. Gas mains broke open and power lines were downed. Fires raged through the city for three days. More than 700 people were killed and quarter of a million were made homeless. The city rose from the destruction relatively quickly, rebuilding within a few years. However, the threat of earthquakes remains as the city sits on the San Andreas Fault, an unstable plate boundary. In 2006, to coincide with the 100-year anniversary, researchers simulated the effects of a similar strength quake. At worst the scenario predicted 3,500 people killed, 130,000 buildings damaged and 700,000 people made homeless. The 1906 earthquake affected a city of 400,000; today there are more than 7 million people in the entire Bay area. The scary truth is that the issue is not *if* the big one happens but *when*.[140]

Rebuilding a city after an earthquake reveals social and political disparity. Diane Davis documents the rebuilding of Mexico City after the earthquake struck on September 19 in 1985.[141] The quake reached 8.1 on the Richter scale. Another, almost comparable quake hit the very next day. The official death toll, probably a significant underestimate, listed 5,000 killed, 14,000 injured and one million residents temporarily made homeless. Hundreds of thousands were made permanently homeless. The greatest damage was in the city center, the administrative heart of the city. The jobs of thousands of public service employees were permanently relocated to other parts of the city. Davis's research of the response reveals a mixed story. While there was rebuilding and reconstruction, with repairs to more than 240 public buildings, many of the private buildings, particularly in the private renting sector, remained unrepaired for years, notably those catering to middle-and lower-income groups. While the government wanted to reassure investors and maintain its legitimacy, citizen groups wanted to recover their city and restore dignity and accountability. The effects of the earthquake revealed violations in building standards and evidence of obvious government corruption. The reconstruction of the buildings went hand in hand with citizens' attempts to rebuild a more democratic city. Environmental disasters revealed cracks in the political system as well as faults in buildings. In Mexico City the earthquake de-legitimized the government, led to a citizens' grassroots movement and the election of a new more democratic mayor. As Davis notes, "The reverberations of the earthquake, in short, were deep and long lasting, and they extended far beyond the built environment to the social and political life of the city."[142]

Responses to earthquakes are not always a route to greater democratization, as they can also reinforce the power of central authorities. On the morning of July 28, 1976, an earthquake that registered 7.8 on the Richter scale hit the Chinese city of Tangshan. Three-quarters of the industrial buildings were destroyed and almost all the residential buildings. The official death toll is listed at 240,000 but may be twice as high. The low-rise industrial city was almost totally obliterated. Within a decade the city was rebuilt. The 1976 Comprehensive Plan imagined a modern city more resistant to earthquakes. Under the very strong and centralized control of the Chinese Communist Party, a more planned city was reconstructed that had more green space and greater land use controls. The reconstruction of the city was "an arena for the display of political authority in the communist regime."[143]

The resilient city

Cities are resilient. Cities have both experienced environmental disasters and transcended them. The term "resilient city" has been coined by urban planners Lawrence Vale and Thomas Campanella.[144] They write of the ability of cities to survive disasters and employ an evolutionary model of recovery consisting of four stages:

Emergency responses: may last from days to weeks and involves rescue; normal activities cease.

Restoration: from 2 to 20 weeks; involves the re-establishment of major urban services and return of refugees.

Replacement and reconstruction: 10 to 200 weeks; the city is returned to pre-disaster levels.

Development Reconstruction: from 100 to 500 weeks; commemoration and betterment.

Into this general picture the experience of particular cities may vary with different equity outcomes. Whose city is destroyed and whose city is commemorated and developed are important questions in assessing the rebuilding of disaster-hit cities. In an interesting study Fallahi, an architect, looked at the rebuilding of the city of Bam, Iran, after the 2003 earthquake that killed 32,000 and leveled the old city. Almost 20,000 homes were destroyed. Temporary shelters were often provided inside the plots of land allowing households to rebuild while also being rehoused. Fallahi makes the point that being homeless does not necessarily mean being buildingless and describes a process of "experts" learning from local people how to rebuild the city. Rebuilding can range from the top down of Tangshan to the more participatory of Bam.[145]

Vale and Campanella suggest a number of axioms of resilience:

- Narratives of resilience are a political necessity
- Disasters reveal the resilience of governments
- Narratives of resilience are always contested
- Local resilience is linked to national resilience
- Resilience is underwritten by outsiders
- Urban rebuilding symbolizes human resilience
- Remembrance drives resilience
- Resilience benefits from the inertia of prior investment
- Resilience exploits the power of place
- Resilience casts opportunism as opportunity

- Resilience, like disaster, is site-specific
- Resilience entails more than rebuilding

Conclusions

Cities are susceptible to environmental hazards. They are subject to droughts and floods, heatwaves and biting cold, earthquakes and volcanic eruptions. The past is littered with examples of cities that were demolished from Pompeii to Lisbon to Galveston. Everyone will remember those images from the tsunami in South East Asia, as waves full of debris and mud swept people up and away in a deadly torrent. The images were a cruel reminder of the unforeseen physical forces that can wreak havoc on our cities and our lives. Disasters are social experiences; they affect poor cities more than rich cities and poor residents more than rich residents. Hazards become disasters though the filter of economic inequality and social injustice.

But cities are also resilient. Resilient cities express the power of hope and opportunity in the face of disaster. Cities have risen, phoenix-like from ashes, a testament to the pulsing life force that is found in cities. The most human of inventions, cities, also express the most human of emotions, hope in the face of adversity.

Guide to Further Reading

Arnold, M. (ed.) (2006) *Natural Disaster Hotspots: Case Studies*, Washington, DC: The World Bank.

Bankoff, G., Ferks, G. and Hilhorst, D. (eds) (2004) *Mapping Vulnerability*, London and Sterling, VA: Earthscan.

Chafe, Z. (2007) "Reducing natural disaster risk in cities", in *State of The World: Our Urban Future*, New York: Norton, pp.112–129.

Cutter, S. L. (ed.) (1994) *Environmental Risks and Hazards,* Englewood Cliffs: Prentice-Hall.

Hoffer, P. C. (2006) *Seven Fires: The Urban Infernos that Reshaped America*, Washington: Public Affairs.

Mesiner Rosen, C. (2003) *Limits of Power: Great Fires and the Process of City Growth in America*, Cambridge: Cambridge University Press.

Mitchell, J. K. (ed.) (1999) *Crucibles of Hazards: Mega-cities and Disasters in Transition*, Shibuya-ku: United Nations University Press.

Pelling, M. (2003) *The Vulnerability of Cities: Natural Disasters and Social Resilience*, London: Earthscan.

Pelling, M. (ed.) (2003) *Natural Disasters and Development in a Globalizing World*, London and New York: Routledge.

Platt, R. (1999) *Disasters and Democracy: The Politics of Extreme Events*, Washington, DC: Island Press.

Vale, L. J. and Campanella, T. J. (eds) (2005) *The Resilient City: How Modern Cities Recover From Disasters*, New York: Oxford University Press.

Winchester, S. (2005) *A Crack in The Edge of The World: America and The Great California Earthquake of 1906*, New York; HarperCollins.

Books on Katrina

Brinkley. D. (2006) *The Great Deluge: Hurricane Katrina, New Orleans and the Mississippi Gulf Coast*, New York: HarperCollins.

Dyson, M. E. (2006) *Come Hell or High Water: Hurricane Katrina and The Color of Disaster*, New York: Basic Civitas.

Hartman, C. and Squires, G. (eds) (2006) *There is No Such Thing as a Natural Disaster,* New York; Routledge.

McQuaid, J. and Schleifstein, M. (2006) *Path of Destruction: The Devastation of New Orleans and The Coming of Age of Superstorms*, New York: Little, Brown.

7 Urban ecology

In recent years an area of study and concern has emerged from the traditionally separate areas of "urban" and "ecology". Most urban studies tended to concentrate on the social side of cities and urban theory has traditionally ignored the ecology of the city, while ecology tends to focus on the less urbanized areas. The science of ecology developed out of a primary concern to understand "natural" processes. The emphasis was on ecosystems with minimal human connections. Between 1995 and 2000, of the 6,157 papers in the nine leading ecological journals only 25 (0.2%) dealt with cities. The early models were developed with primary reference to pristine ecosystems. In recent years, however, an urban ecology has emerged that considers cities as sites of biophysical processes interwoven with social processes in complex webs of human-environmental relations. We will use the term urban ecology to refer to the ecology of cities rather than simply the study of ecology in cities.[146] Specific journals such as *Urban Ecosystems* and *Urban Ecology* publish a wide variety of papers in this new and exciting field. Researchers in urban ecology are developing both general models and specific case studies. Mary Cadenasso and her colleagues, for example, theorize the Baltimore metro area as a complex of biophysical, social and built components. They apply standard ecological approaches such as ecosystem, watershed and patch dynamics in order to answer three general questions: what is the overall structure of the urban ecosystem; what are the fluxes of energy, matter, population and capital in the system; what is the nature of the feedback between ecological information and environmental quality. [147] In a more specific vein, Georgi and Zafiriadis look at the impact of trees in parks on urban microclimates. They show, with reference to Thessalonica in Greece, that trees reduced summer temperatures, increased relative humidity and have a general cooling effect.[148]

In the rest of this chapter we will look at some of the main features of urban ecology by considering energy flows, urban footprints, natural capital, biophysical cycles and cities as biotic communities.

Urban Gradients

Ecologists often use urban gradients to investigate sample sites. Sampling sites along the gradient allow researchers to estimate the effects of urbanization. Here is a gradient proposed by Marzluff et al. It is a very coarse-grained scale and finer-grained divisions are necessary for more locally based studies.

Term	% built up	Building density	Population density
Wildland	0-2	0	< 1/ha
Rural/exurban	5-20	< 2.5/ha	1-10/ha
Suburban	30-50	2.5-10/ha	> 10/ha
Urban	> 50	> 10/ha	> 10/ha

Source: Marzluff, J. M. , Bowman, R. and Donnely, R. (eds) (2001) *Avian Ecology in An Urbanizing World*, Norwell: Kluwer Academic.

Cities as flows of energy

Ecologists often conceptualize ecosystems as a set of flows. One of the most important is the flow of energy. Imagine a meadow. The grass and flowers turn the energy of the sun into plant material through photosynthesis. The animals that feed on the plants are producers. Herbivores are primary producers while the animals that eat the animals are secondary consumers. Energy flows through the different levels in a pyramid of trophic levels.

Cities can also be modeled as flows of energy. There are the flows of energy that power a city. We can picture the city as a place where the different energy sources – human muscle power, electricity, nuclear, and wind – provide the basic energy for heat, light, power and transport. Different cities use different energy sources over time (heavier reliance on human power in the ancient cities) and space (such locally sensitive sources as the windmills of Holland). Over the long term of urban history there has been a major shift from human muscle power to mechanical power. There is now a reliance on fossil fuels as a source of energy to power cities. The economies of cities have waxed and waned as energy sources rise and fall. From the coal-field towns of industrial Britain to the oil-thirsty cities of the contemporary world, cities have been shaped by the mix of energy sources they use. The widespread use of energy sources has extended the range of urban living. Large cities can develop in cold climes as well as hot, dry climes.

Energy use varies by the location and wealth of cities. Car-dominated cities in the very cold or very hot climates of rich countries, for example, will tend to use more energy. Cities in rich countries use more energy than cities in poor countries.

Focusing on cities as circuits of energy brings into sharp relief the issue of the efficiency of cities. We can picture a more ecologically informed urban studies developing around critical energy audits of different types of cities and a more urban informed ecology developing new models of fuel efficiency. Numerous questions are prompted by such an approach: for example, does public transport use less energy than private transport? Are certain building forms and spatial arrangements of the built form more conducive to energy efficiency than others? When walking or bicycling replaces auto transport – albeit only possible for short distances – are there other benefits in addition to reducing energy inputs? Do people who walk more tend on average to be healthier and fitter and thus require less health care expenditure?

Food is also an energy source. Imagining the city as flows of food highlights the way cities interact with the humanized agricultural landscapes. Tracing the daily consumption of calories of urban residents from the purchase in retail establishments through wholesale distributors to primary producers would focus attention on the web of connections that span across time and space and link changing agricultural practices with the changing food demands of city dwellers. These food networks are mediated by complex factors of economy, culture and society. Following the interconnecting physical and social nexus that links even the simplest act, such as purchasing a morning coffee, would allow us to see how urban consumers are linked in complex relations with distributors and producers. The modeling of these food pathways and their changing configuration, size and character is a novel way to understand people – environment relations in linkages that connect ecological processes with business cycles and changing patterns of food demand and supply.

The supply of food also embodies social inequalities. Researchers have used the environmental term desert to refer to areas of the city with limited access to nutritional food. Neil Wrigley's work on "food deserts" in British cities shows that these deserts reflect social exclusions and are an important element in health inequalities.[149]

Urban footprints

Ecologists have developed the notion of the ecological footprint to refer to the total area of productive land required to support an ecosystem. The ecological footprint measures how much land and water area a human population requires to produce the resources it consumes and to absorb its wastes.[150] The concept has been used at the global and individual scale as well as at the level of an urban region. The ecological footprint of an urban region includes all the land necessary to support the resource demands and waste products of a city. An early study of the ecological footprint of the Vancouver region suggested that it needed 19 times the surface area of the home region to satisfy the need for food, clothing, energy and shelter. Energy needs cast the biggest footprint.[151] Figure 7.1 shows the urban footprint for London.

Footprints vary. In Las Vegas, for example, the explosive growth of the city, with more than 5,000 people moving in every month, has increased the demand for water. The city relies on the Colorado River for much of its water but the amount is fixed. The city is planning to build a pipeline 250 miles to the north to tap into underground water. But this is the same water source that ranchers and farmers use. Environmental activists argue that this is creating "a sacrifice zone of thousands of square miles".[152]

Measuring the ecological footprint of a city, allows us to fine-tune concepts such as ecological overshoot and changing carrying capacity, and give a better measurement to the idea of long-term sustainability. And developing the idea of a city's ecological footprint will necessitate a more nuanced ecological accountancy than the simple models used for either individual households or for the total global populations. There are collective goods and services in cities that can combine with individual patterns of consumption to both maximize and minimize the

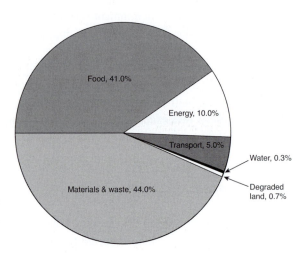

Figure 7.1 London's ecological footprint
Source: Adapted from City Limits Executive Summary, accessed at http://www.citylimitslondon.com

collective footprint. For example, a city that is more reliant on private autos than public transport has greater energy needs and thus has a larger footprint. In cities, individual and national footprints combine and interact to produce distinctly urban regional effects.

Rethinking New York city's footprint

According to Edward Glaeser, professor of economics at Harvard,

> Manhattan, not suburbia, is the real friend of the environment. Those alleged nature lovers who live on multi-acre estates surrounded by trees and lawn consume vast amounts of space and energy. If the environmental footprint of the average suburban home is a size 15 hiking boot, the environmental footprint of a New York apartment is a stiletto-heeled Jimmy Choo. Eight million New Yorkers use only 301 square miles, which comes to less than one-fortieth of an acre a person. Even supposedly green Portland, Oregon is using up more than six times as much land a person than New York.

> New York's biggest environmental contribution lies in the fact that less than one-third of New Yorkers drive to work. Nationwide, more than seven out of eight commuters drive. More than one-third of all the public transportation commuters in America live in the five boroughs. The absence of cars leads Matthew Kahn, in his fascinating book, "Green Cities," to estimate that New York has by a wide margin the least gas usage per capita of all American metropolitan areas. The Department of Energy data confirm that New York State's energy consumption is next to last in the country because of New York City.

> ...When Manhattan builds up, instead of Las Vegas building out, we are saving gas and protecting land. Every new skyscraper in Manhattan is a strike against global warming. Every new residential high rise means a few less barrels of oil bought from less than friendly nations belonging to the Organization of the Petroleum Exporting Countries.

> ...The great problem with being reflexively anti-growth is that development in America is close to being a zero-sum game. New homes are going to be built to meet the needs of a growing population. If you stop development in some areas, you are ensuring more development elsewhere. A failure to develop New York means more homes on the exurban edges of America.

Source: Glaeser, E. (2007) "The greenness of cities", New York Sun, January 30. Accessed at: http://www.nysun.com/article/47626, retrieved April 6, 2007.

Cities have vast footprints that encompass not only the immediate surrounding area but extend across the nation and the globe. Tracing the footprint of an individual city through time, with different types of cities, would be a fascinating way to connect urban history with ecology.

Capital

In recent years ecologists have used the term ecosystem capital or natural capital.[153] The term is employed to highlight the human use of ecosystems as goods and services that have value. Ecosystem capital is a source of wealth and income generation. Employing the term capital in this way highlights the economic use of ecosystems, and the importance of protecting ecosystem capital. Take one example: we can see the tropical rainforest as an ecosystem capital that provides goods in the form of wood products as well as the services involved in tourism. To overexploit the wood products would destroy the tourist income.

The term 'capital' involves more than an economic accountancy model of the physical environment. Capital is above all a social relationship. Studies of urban ecologies would be enlivened and propelled by a more critical theoretical view of the physical environment as a social relationship between different groups as mediated through the medium of capital. To use the term capital is not simply to provide a measurement for the rational use of resources but also the basis for a critical take on the social use of resources.

Biophysical cycles and social processes

An important theme in traditional ecology is the modeling of biogeochemical cycles. Three important cycles have been recognized; the carbon cycle, phosphorus cycle and nitrogen cycle. These cycles all share a similar characteristic; they are subject to the basic Newtonian principle that matter can neither be created nor destroyed. The cycles are a closed system of fixed stock.

Carbon atoms exist as carbon dioxide in the air and bicarbonate in water. Carbon is taken up by plants through photosynthesis and metabolism. The human intervention in the carbon cycle is most obvious in the burning of fossil fuels. By burning the fossil fuels necessary to power cities we have intervened dramatically in the carbon cycle, adding more CO_2 to the air. Levels of carbon dioxide are at historic levels and are responsible for the warming of the planet. By reducing fossil fuel consumption, using energy more efficiently, developing new forms of energy including nuclear, tide, wind or others we can reduce the level of fossil fuel emissions.

Cities also play less obvious roles in the carbon cycle. Jenkins and colleagues show the importance of vegetation in cities and urban soils as important storage sites for carbon, while Pouyat and his colleagues show that urban soils, especially in residential areas, sequester large amounts of soil organic carbon.[154]

The phosphorus cycle is a mineral nutrient cycle. Phosphate in rocks and minerals is taken up by plants, then moves through the food cycle. The humanly modified phosphorus cycle involves the mining of the mineral and its application as fertilizer and use as detergent. Current food production is predicated upon heavy applications of phosphorus to the soil. The agriculture that feeds cities around the world has been reliant on heavy application of phosphorus, which leaches into water. Over-fertilized water has resulted in algae blooms that reduce the amount of light in streams, lakes and other water bodies. Cities play an important role in this process, not only because their heavy food demands have prompted greater use of phosphorus but also because the lawns and gardens of towns and cities are in part created by the enhanced application of nutrient supplements that contain phosphorus. The lush green lawns of suburbia are engineered by heavy applications of phosphorus and bound up in a lawn-chemical economy and its environmental impacts.[155]

Nitrogen is found in the air. Bacteria convert this nitrogen into ammonium in a process known as nitrogen fixing. Legumes are important nitrogen fixers. Plants that draw on the nitrogen found in mineral form, then pass through the usual route of herbivores and carnivores to decomposers. Humans also fix nitrogen. The Haber-Bosch process converts nitrogen gas into ammonia. In the combustion of fossil fuels nitrogen is oxidized into nitrogen oxides. The rate of nitrogen fixing has increased dramatically because of the cultivation of legumes such as beans, soybeans and alfalfa, the application of fertilizers rich in nitrogen and the burning of fossil fuels. Excess nitrogen is leached into the soil and is the primary cause of dead zones in rivers and oceans. Nitrogen cycling is altered by urbanization and land use change. Urban and suburban watersheds have much higher rates of nitrogen loss than forested watersheds.[156]

This very brief view shows how cities play an important role in modifying biogeochemical cycles. We live in a world of modified biogeochemical cycles. Early models rarely included human biogeochemical controls such as impervious surface proliferation, engineered aqueous flow paths, landscaping choices and differing levels of human population density, size and growth rates.

New models are being developed that try to capture the complex human–environment relationship of biogeochemical cycles in cities.

There is a complex relationship between biophysical processes and socioeconomic behaviors. If we consider individual plots of urban land, for example, it is clear that the knowledge of physical processes influences market choice and policy. To take an obvious example, flood-prone land will be avoided by developers or protected by legislation. Market choices such as the decision to develop certain pieces of land will be determined in part by the policy framework and in part by the opportunities and constraints of the local biophysical processes. In other words, there is a complex system of decisions that links ecology and economy, society and policy. Land use models are being improved by incorporating a more explicit understanding of biophysical processes, while ecological models are enhanced by a more explicit consideration of socioeconomic processes.

Long-term studies of urban ecosystems

The US National Science Foundation funds long-term ecological research projects at 24 sites in a Long-Term Ecological Research Network. Only two of them specifically study the ecology of urban systems. The two sites, one in Baltimore MD, the other in Phoenix AZ, provide information on urban ecological processes and trends. The program brings together researchers from the biological, physical and social sciences. The Baltimore Ecosystem Study (BES) aims to understand metropolitan Baltimore as an ecological system. The Central Arizona–Phoenix Long-Term Ecological Research focuses on a metropolitan region in an arid-land ecosystem and research examines the effects of urban development on the Sonoran Desert and the impact of ecological conditions on urban development.

Visit these websites for updates on research projects: Baltimore Ecosystem Study: http://www.beslter.org/ Central Arizona–Phoenix Long-Term Ecological Research: http://caplter.asu.edu/

Cities as biotic communities

An important element of traditional ecology is the idea that ecosystems are biotic communities of plants and animals. We can begin to think of urban ecosystems as a distinctive ecological category rather than merely as disturbance sites to be unfavorably compared to pristine sites. Urban development initially increases local extinction rates and rates of loss of native species. The general tendency is for the replacement of native species by non-native species. However, mature urban environments create a rich range of different ecologies and habitats. Cities contain open space as well as built-up areas with sites varying in vegetation cover and habitat variety. The city is more accurately described as an ecological mosaic than a single category.

A more dynamic perspective that sees change as a constant has replaced the old idea of ecosystems as a set of successions leading to a stable climax. Ecological climaxes are always temporary and sometimes short lived. It is more appropriate to consider ecosystems as ensembles of plant and animal species that are constantly adapting to small-scale and large-scale change over short and long periods of time. The precise form of change depends on a range of environmental factors. Changes in environment can cause shifts in populations of particular species.

The city provides opportunities and constraints to different species. Foxes in Europe and coyotes in the US have shown themselves remarkably adaptable to urban environments. Species that have a wide distribution, a large population, a high degree of genetic variation and a large number of offspring will tend to survive and adapt. Cities are bringing about discernible changes in biotic communities that favor some species rather than others. The urban mosaic provides a variety of ecological niches. Blewett and Marzluff, for example, looked at the distribution of cavity-nesting birds in Seattle, WA. They found that suburban landscapes with highly interspersed land cover had higher densities of black-capped chickadees, nuthatches, flickers and downy woodpeckers while suburban landscapes with more forest cover had higher densities of creepers, chestnut-backed chickadees, pileated and hairy woodpeckers.[157] In a study of birds in California and Ohio, Blair found species richness and diversity was highest in residential areas compared to preserves. In other words the mosaic of habitats in suburban areas resulted in more species than many designated preserves. This study suggests that birds that are urban exploiters successfully reproduce, invade locally, and have multiple broods while urban avoiders do not reproduce.[158]

We can identify specific urban biotic communities. Plants in cities, for example, are exposed to more pollutants, warmer temperatures and higher levels of carbon dioxide and nitrogen deposition than plants in comparable rural areas. Certain tree species can tolerate and survive heavy air pollution. The London Plane tree, for example, thrives in the polluted air of London. Similarly, in the Northeast US, maple and crab apple trees are more tolerant of air pollution than many other species. Gregg and colleagues grew the same cottonwood clone in urban and rural sites. They found that the biomass in urban areas was double compared to rural sites. The principal reason was that while ozone reduced plant growth, the higher levels of nitrous oxides in the urban sites suppressed ozone levels.[159]

Cities are engineered landscapes. Urban plant communities, for example, now reflect socioeconomic and cultural factors as well as the traditional ecological factors. Cultural preferences are interconnected with the physical factors. In addition to elevation and land use history, factors such as family income and housing now explain plant diversity. Along with biotic factors, human variables such as household preferences and family wealth structure plant community formation in residential areas.[160]

Urban biotic communities can also be engineered. It is now well recognized that planting trees and shrubs around buildings can reduce summer heat. A study of the Chinese city of Harbin compared the dust removal capacities of 28 tree species. The results showed that *Pica koraiensis* and *Juniperus rigida* were the ideal conifers for dust removal, and *Populus alba x berolinensis*, *Lonicera maackii* and *Prunus maackii* were the ideal deciduous tree species. These tree species have deep channels of dense hair on their leaf surface that is more effective in dust removal. The authors proposed that dust removal capacity be an important basis in the choice of tree planting in the city.[161]

Reimagining the city as a biotic community will allow us to explore the idea that the city is an important ecological category where plants and animals live and where new ecological niches are being opened and closed as urbanization increases and changes. Urbanization has transformed ecosystems around the world and has encouraged the development of new forms of more urban-tolerant plant, animal and bird life.

Ecologies of a model city region

Using the models of ecology is not new in urban studies. In the first third of the twentieth century the Chicago School theorists, Robert Park and E. W. Burgess, developed models of city structure and social change that drew implicitly on the ecological language of invasion and succession. They used biological terms to describe the city and the processes of social differentiation over space and time. While later critics have accused them of a crude social Darwinism, their biological referencing was more rhetorical and metaphorical than causal. Drawing on their work we can suggest the following simple urban model of the city as consisting of a central business district, zones of transition, a ring of inner city residences tending toward lower density and a suburban frontier where non-urban land is being transformed into urban land uses; we then have a simple structure for suggesting important urban ecological zonal processes.[162]

The most significant feature of the central business district is the dense collection of buildings and the agglomeration of artificial surfaces and urban land uses. Here, because of the density of impermeable surfaces, flooding is at its most pronounced and the urban heat island is the hottest. Flooding is only contained by elaborate artificial drainage systems that can deal with the abrupt peaking of water discharges. The urban heat island is the distinct warming of urban areas compared to rural areas. In the average US metro region, for example, urban areas are 2° to 10° C warmer than surrounding rural areas. The heat island is reduced by open space and water bodies and heightened by urban building. Temperatures increase toward the dense city center not only because tree and vegetation have been cleared but also because tall buildings and narrow streets restrict air flow and lead to the heating of trapped air. There are in fact two heat islands: a canopy layer at the ground level and a higher boundary layer heat island. A mapping of the canopy surface temperature in the city of Baltimore in October 2001 showed that there was an 8° C difference between local forested areas and the urban areas with a major heat island clearly identifiable in the city center. In the city center there was an island of cooler temperatures caused by the canyon shading of high-rise buildings. In these shadowed city canyons temperatures were 5–10° C cooler than the rest of the city.[163]

The urban heat island can be lessened by covering the roof of buildings with vegetation to reduce heating while also improving air quality. The city of Chicago, for example, has an active policy of encouraging the vegetation of commercial roofs through grants, tax breaks and other financial inducements to developers and builders.

The city is in the process of continual change. The Burgess model assumed only one zone of transition, an area of housing deterioration. We can identify a variety of zones of transition including the abandonment of older housing, the deindustrialization of industrial factory zones, and areas of new building and gentrification. In each of these zones new urban ecologies are being created. Take the case of the former factory areas. The decline of manufacturing especially in the older inner city areas is creating the brownfields of an industrial legacy with issues of soil contamination and pollution. The trajectory from factories to brownfields to greenfield sites is not only a social process but also an ecological transformation.

Within the residential areas there are a variety of ecological processes. Expanding cities leave in their wake fragments of natural vegetation and also create new ecologies. One of the most obvious is the creation of engineered green spaces of gardens. On the down side this can involve the application of fertilizers creating increased levels of nitrogen and phosphorus in the soil and water runoff. In many cities there is a move for more "natural" gardens especially in dry and arid climates where the lush green lawn is only possible by constant applications of water and fertilizer. But on the more positive side, city gardens can also protect threatened species. Many city gardeners consciously grow plants that attract threatened bird and insect life. The role of domestic gardens in ecological transformation and the creation of new ecological niches has long been overlooked. Suburban areas are complex systems of built form and green space that can contain refuges for a whole range of plant and animal life.[164]

At the edge of the city the dominant ecological change is the creation of the built form across formerly non-urban land uses. A 200-year story of changing land use in the Baltimore Washington corridor is depicted at http://biology.usgs.gov/luhna/hinzman.html The pace of change picks up dramatically in the latter half of the twentieth century. At http://pubs.usgs.gov/circ/2004/circ1252/2.html sprawl in the Boston area from 1973 to 1992 is clearly visible as red areas of developed land spread around the coastline and extend inland, from 515 square miles to 764 square miles. More recent changes are recorded by satellite-based land-surface mapping techniques. The images produced by the Mid-Atlantic Regional Earth Science Applications Center (RESAC) show clear evidence of dramatic land use change in the Baltimore–Washington corridor (http://www.geog.umd.edu/resac/research.htm). From 1986 to 2000 the amount of developed land increased from around 600,000 acres to over a million acres. The RESAC satellite imagery reveals that almost 20 percent of the total land surface is now devoted to developed land and at current rates this will increase to 1.8 million acres or 36 percent of the region's land surface by 2030. Sprawl provides some benefits. It gives middle-income households access to a wide-range of safe, affordable homes in tight housing markets. A wide scatter of employment, retail and commercial development spreads economic opportunities throughout the metropolis. But there are costs to sprawl. The first is the reliance on oil-based private transport. Sprawl is a form of development that is too low density to support public transport. The heavy and in some cases total reliance on private auto transport imposes a heavy environmental price in terms of air pollution, and the increasing dedication of space for roads and parking. The reliance of a built form precariously balanced on one fossil fuel with large and fluctuating costs raises issues of sustainability and affordability.[165] The transformation of mixed land use into a suburban landscape can result in the loss of species habitat, and even when green spaces are kept they are often as isolated islands that are unable to function as true niche ecologies.

Urban development consists of the replacement of permeable land with impermeable surfaces. More paved surfaces means less area for water to drain into soils, thus bringing more water into fewer drainage systems, increasing erosive power and sediment load. The replacement of permeable land with impermeable tarmac creates problems from more flooding to increased runoff that collects toxins in the groundwater. A study of non-tidal streams in Maryland, US found that when 10 percent to 15 percent of areas are paved then increased sediment and chemical pollutants reduce water quality; at 15 percent to 20 percent there is markedly reduced oxygen levels in streams; and at 25 percent many organisms die. Many studies decisively document the local impairment of streams with an increase of urban land use.[166]

Conclusions

Even this very brief description of exemplary ecological processes in the different zones of the city reveals the possibility of new models of cities, models that incorporate both biophysical and socioeconomic processes. Integrating the ecology into urban models and the urban into ecological models will enrich both areas of traditional study.

Guide to Further Reading

Goudie, A. (2005, 6th edn) *The Human Impact on The Natural Environment,* Oxford: Blackwell.

Wright, R. T. (2005, 9th edn) *Environmental Science: Towards a Sustainable Future,* Upper Saddle River, NJ: Pearson–Prentice Hall.

Platt, R., Rowntree, R. A. and Muick, P. C. (eds) (1994) *The Ecological City*, Amherst: University of Massachusetts Press.

Alberti, M. (2005) "The effects of urban patterns in ecosystem function", *International Regional Science Review* 28: 168–199.

Collins, J. P., Kinzing, N. B., Grimm, W., Fagan, D., Hope, J. Wu and Borer, E. T. (2000) "A new urban ecology: modeling human communities as integral parts of ecosystems poses special problems for the development and testing of ecological theory", *American Scientist* 88: 416–425.

Kaye J. P., Groffman, P. M., Grimm, N. B., Baker, L. A. and Pouyat, R. V. (2006) "A distinct urban biogeochemistry?" *Trends in Ecology and Evolution* 21: 192–199.

On urban footprints

The EPA has posted interactive "footprints" for several US cities. You can view them at: http://www.epa.gov/watertrain/smartgrowth/02animation/chspk.htm

8 Water pollution and the city

Water is essential to urban life. In cities there are two main water issues, the contamination of water sources and ensuring the supply of clean water. In this chapter we will focus both issues. At the outset we will examine the case of water pollution in US because it has some of the most well-developed clean water regulations in the developed world. We will also highlight water supply and management issues in cities in the developing world.

Wastewater in US cities

In the US, as in other rapidly industrializing-urbanizing nations, one of the major reforms to the "industrial city" of the nineteenth century was the establishment of sewer systems to collect wastewater. While many cities built wastewater treatment facilities in the nineteenth and early twentieth centuries, population growth meant that by the middle of the twentieth century, the volume of sewage and stormwater exceeded the processing ability of most treatment plants. This was particularly noticeable during heavy rains.

Combined Sewage Overflow (CSO) refers to the temporary direct discharge of untreated water. CSOs occur most frequently when a city has a combined sewer system (CSS) that collects wastewater, sanitary wastewater and stormwater runoff in various branches of pipes, which then flow into a single treatment facility (see Figure 8.1). This type of sewer system is prevalent in the urban centers in the Northeast and Great Lakes regions. CSSs serve approximately 772 municipalities with approximately 45 million people. During dry weather, CSSs transport wastewater directly to the sewage treatment plant. However, rainwater or urban storm runoff is not directed separately, but co-mingles with household wastes and industrial wastes. When it rains, few facilities can handle the sudden increase in volume of water and, as a result, the excess volume of sewage, clean water and stormwater is discharged untreated into rivers, lakes, tributaries and oceans. CSOs contain

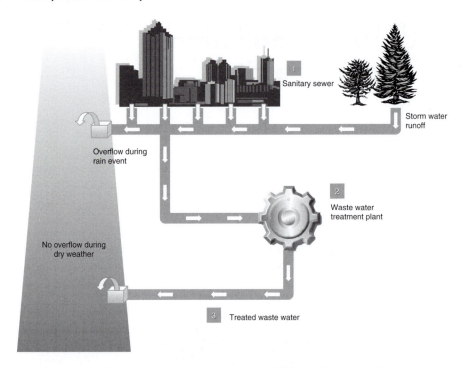

Sanitary sewer

Storm water runoff

Overflow during rain event

1

Waste water treatment plant

2

No overflow during dry weather

3 Treated waste water

Figure 8.1 *Combined Sewer Systems. In cities with combined sewer systems, stormwater runoff from rain combines with sewage in the same pipe system and is discharged directly into river, creeks or estuaries prior to any treatment. The discharge consists of many pollutants, untreated sewage and debris.*
Source: Lisa Benton-Short

not only stormwater but also untreated human and industrial waste, toxic materials and debris. CSOs are among the major sources responsible for beach closings and shellfish restrictions and the contamination of drinking water. CSOs caused some 150 beach closures from 1999–2002. Table 8.1 lists the CSO releases for selected US cities. Residents of these cities are often warned to avoid contact with river water or beach water for several days after periods of heavy rainfall.

The other major type of public sewer system is the Sanitary Sewer Systems (SSSs), built in cities since the start of the twentieth century. SSS systems have separate pipes that collect sewage and stormwater separately. However, when it rains the excess stormwater can overload the system and so stormwater is discharged untreated (see Figure 8.2). Recent studies have found that during the initial rainfall, the concentration of pollutants in urban stormwater rivals and in

Table 8.1 *Combined Sewer Overflow (CSOs) releases, 1999*

City	Billions of gallons
Chicago	27.2
Atlanta	5.3
New York	84.5
Philadelphia	20.4
San Francisco	1.7
Washington DC	2.2
Richmond	4.1
Cleveland	5.9

Source: Adapted from Environmental Protection Agency (2006) *Report to Congress on the Impacts and Control of CSOs and SSOs.* Executive Summary

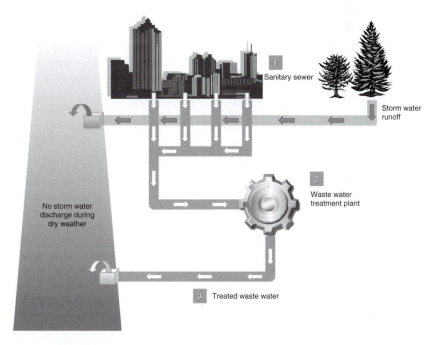

Figure 8.2 *Separate Sanitary Systems. For cities that have separate sanitary and stormwater systems, stormwater runoff during rains are collected separately from sewage. The stormwater runoff, which contains pollutants and debris, is diverted from the wastewater treatment plant and discharged directly into rivers, creeks or estuaries.*
Source: Lisa Benton-Short

some cases exceeds sewage plants and large factories as a source of damaging pollutants, beach closures and shellfish decline.[167]

An EPA study conducted from 2001–3 found that each year CSO events discharge 850 billion gallons and SSOs discharge another 10 billion gallons per year.[168] Pollutants associated with CSOs and SSOs and other municipal discharges include nutrients (which can stimulate the growth of algae that deplete dissolved oxygen in surface water, thereby "asphyxiating" fish), bacteria and other pathogens (which may impair drinking water or disrupt recreational uses) and metals and toxic chemicals from industrial and commercial activities and households.

Point and nonpoint pollutants

There are two main sources of water pollution, point and nonpoint. Point sources are those where there is a clear discharge mechanism such as effluent pipes or outfalls. According to the US Environmental Protection Agency, point sources are defined as any "discernible confined and discrete conveyance including but not limited to any pipe, ditch, channel, tunnel, conduit, well, discrete fissure, container, rolling stock, or concentrated animal feeding operation, or vessel or other floating craft from which pollutants are or may be discharged." These stationary devices can be measured for the amount of pollution discharged. The main point sources of pollution in cities are industrial and municipal facilities. Point sources of water pollution are relatively easy to monitor and control. The National Pollution Discharge Elimination System is a permit program of the Clean Water Act. As a result, point source pollution in the US has declined significantly since the implementation of the clean water legislation.

Nonpoint sources are any source from which pollution is discharged which is not identified as a point source. There are two main types of nonpoint sources of water pollution in cities. The first is urban runoff, a term that refers to the various pollutants that accumulate in soil and on roadways that are washed into the sewage systems during floods or rains. As rainwater makes it journey over roads, parking lots and other urban structures, it picks up a variety of pollutants. The EPA notes that urban runoff is now the largest source of pollution to estuaries and beaches. The second type of nonpoint include wastes and sewage from residential areas. These wastes are collected in the larger sewer system, and it is impossible to tell their origin. These pollutants are harder to regulate and reduce. In addition, nonpoint sources are often intermittent and diffuse, making it hard to quantify individual contributions. The Clean Water Act 1997 Amendments,

Continued

Section 319, requires states and territories to develop programs to deal with non-point source pollution. The EPA has set aside $370 million in funds to implement nonpoint source pollution controls.

Point Sources (primarily industrial wastes)	Nonpoint Sources
Oil and grease	Salt
Heavy metals	Oil
Organic chemicals	Gasoline
Acids and alkalides	Antifreeze
Salts	Floatables and Debris (plastics, cans, bottles)
Solvents	Pesticides and Fertilizers
Organic matter	Organic Matter (including human and animal wastes)
Suspended solids	Microbial pathogens
Heat/thermal	Solids and sediment
pollution	Sewage
	Heavy Metals (chromium, copper, lead and zinc)

Sources: US EPA (2007) "Polluted runoff (nonpoint source pollution: managing urban runoff") Document EPA841-F-96-004G retrieved at www.epa.gov/owow/nps/facts/point7.htm March 2007, and EPA (2006) "Polluted runoff: nonpoint source pollution: the nation's largest water quality problem" retrieved at www.epa.gov/owow/nps/facts March 2007.

Once sewage has reached a treatment plant, physical process can separate biological solids from wastewater. Screens or filters catch raw sewage and debris, channeling them to sludge ponds where the organic materials are broken down by bacteria. Often sludge that has dried is shipped to landfills.

By the early twentieth century, some cities began to install secondary treatments in their facilities. Secondary treatment involves aeration and finer filtration. It is a more controlled way of producing bacterial growth to break down biological solids. It is estimated that secondary treatment eliminates more than 95 percent of disease-carrying bacteria from the water. Since 1972, Congress has provided $69 billion to assist cities in constructing secondary wastewater treatment plants.[169] State and local governments have spent more than $25 billion. Yet it is also estimated that cities require another $140 billion to achieve the goals established.

Water quality legislation

In the United States, there were a series of legislative precedents for water quality including the 1948 Federal Water Pollution Control Act, the 1956 Water Pollution Control Act, and the Water Quality Act of 1965. These were among the first comprehensive statements of federal interest in clean water, but mounting frustration over the slow place of pollution cleanup efforts along with increased public interest in environmental protection set the stage for the 1972 amendments. Most environmental experts point to the 1972 Federal Water Pollution Control Act Amendments (commonly called the 1972 Clean Water Act) as setting the framework for the last 35 years of water pollution policy. It was remarkable for several reasons. It is the principal law governing pollution of the nation's surface waters. For the first time, a federal agency was in charge of water pollution, taking some control from the states and establishing basic federal standards, with which all states had to comply. The Clean Water Act (CWA) of 1972 required that all states set water quality standards, something that had been previously left to each state to decide. The CWA also provided technical tools and some financial assistance to address the many causes of water pollution. The CWA was highly ambitious: its major goals were to eliminate the discharge of water pollutants, restore water to "fishable and swimable" levels, and completely eliminate all toxic pollutants. Today the three ambitious goals of the CWA of 1972 have not been achieved, yet they continue to provide the framework for how the federal government (and municipalities) deal with water pollution.

Two major components of the CWA impact cities directly. Title II and Title VI authorize federal financial assistance for municipal sewage treatment plant construction and set out regulatory requirements that apply to industrial and municipal dischargers. Both of these are critical to understanding the relationship between cities and water pollution reform. With federal assistance, many cities were able to upgrade sewage treatment plants, adding secondary or tertiary treatment, or increasing the volume the treatment plants could process. Title VI, which applied mainly to industrial and municipal discharge, resulted in monitoring and management programs for point source pollution. The result has been a decrease in industrial point pollution.

Assessing the CWA

We now have a more complete picture of the volume and types of municipal pollution. Thirty-five years after the initial Clean Water Act of 1972, the results are mixed. Point source pollution from industrial sources has been reduced. In many rivers and lakes, oxygen levels have recovered due to the filtering out of organic wastes. Some pollutants have declined, but others are on the increase. In 1970, more than 66 percent of rivers, lakes and estuaries were too polluted for fishing or swimming. By the late 1990s, 40 percent remained too polluted. But Lake Erie, once "dead" has recovered. And in the Cuyahoga River, once a stark symbol of the plight of America's rivers, the blue herons have returned, and the city now boasts new marinas lining a river walk and up-market sidewalk cafes. In 1970, only 85 million Americans had a wastewater treatment plant. By 2000, 173 million did (more than 70 percent of the population).[170] And yet more than 15 large cities continue to discharge some 850 billions of gallons of untreated sewage and runoff each year during heavy rains and floods (see again Table 8.1). In some cities, beach closures have increased in places where the combined sewers are older and inadequate for the population. And while the total releases of toxic pollutants have decreased (particularly with regard to mercury and DDT), more than 47 states still post fish consumption advisories for contamination due to mercury, PCB, dioxins and DDT.

The mixed results reflect several factors. First, new technologies impact on our ability to measure pollution in smaller and smaller amounts. Scientists are able to document levels of pollution in parts per million or billion, levels that we were unable to detect previously. In addition, our knowledge and understanding of the impact of pollution on human health and environmental health is improving, and this adds new "pollutants" to the list and can require a readjustment of allowable exposure standards. There are now approximately 80,000 chemicals registered with the EPA: few have been comprehensively tested, many resist breakdown and some accumulate in fat tissues. Another factor is that the EPA has tended to focus on "end of the pipe" solutions—that is regulating discharges at the source of the polluter rather than focusing on ways to encourage "front end" controls or preventive measures. The EPA has been less successful at encouraging the reduction of waste in the first place. While millions of dollars have been available to assist in the development and adoption of technological solutions, less is available for programs that encourage reduction or reusing of pollutants in the first place.

In 2000, the EPA established a set of federal policies and incentives for reducing urban storm runoff and to help smaller cities (of under 100,000 population) develop adequate wastewater treatment facilities. In addition, the 1987 Clean Water Act amendments directed states to develop and implement nonpoint pollution management programs. Under section 319 of the Act, $400 million in grants was made available for states to assess the extent of nonpoint source water quality impairments and to develop and implement plans for managing nonpoint sources.

The legacy of the Cuyahoga River fire

In June 1969, the Cuyahoga River in the city of Cleveland caught fire. The river, filled with kerosene and other flammable material was probably ignited by a passing train that provided the spark. Although it burned for only thirty minutes, the incident and the famous photograph of the river on fire became a pivotal part of an emerging environmental movement. The Cuyahoga became the symbol of urban water pollution and the need for the federal government to become involved in cleanup and regulation. Ironically, the photograph associated with the 1969 fire was actually from an earlier fire in 1952. The Cuyahoga had a history of fires, but only the fire in 1969 drew national attention. Below, Jonathan H. Adler, Associate Professor of Law at Case Western Reserve University remarks on the interesting legacy of the Cuyahoga:

> The Cuyahoga fire was a powerful symbol of a planet in disrepair and an ever-deepening environmental crisis, and it remains so to this day. That a river could become so polluted as to ignite proved the need for federal environmental regulation. Following on the heels of several best-selling books warning of ecological apocalypse and other high-profile events such as the Santa Barbara oil spill, the 1969 Cuyahoga fire spurred efforts to enact sweeping federal environmental legislation. The burning river mobilized the nation and became a rallying point for the passage of the Clean Water Act.

> Why didn't states act earlier? In the 1950s, let alone in 1910 or the 1930s, environmental issues did not yet rank with concerns for economic development, technological progress and other social ills. Policymakers at all levels of government knew little about the health effects of pollution and paid it little heed. While the environmental problems that plagued Cleveland and other parts of the nation are obvious in hindsight, the nature and extent of these problems were not always readily apparent at the time. Once the demand for greater pollution control emerged, action began.

> The 1969 fire was not evidence of nationally declining water quality either. To the contrary, early cleanup efforts on the Cuyahoga appear representative of state and local efforts nationwide. Throughout the 1950s and 1960s states began to recognize the importance of environmental quality and adopted first-generation environmental controls. As the nation became wealthier, and the knowledge base improved, attention to environmental matters increased. It is well established that wealthier societies place greater importance on

Continued

environmental protection. They also have greater means to protect environmental values. ... And contrary to the common fable, in most cases state and local governments were the first to act. The 1969 fire was a catalyst for change because it was the wrong event at the right time. It was neither an impressive fire, nor one with a significant ecological impact. It may have brought greater attention to the serious environmental problems of the time, but it did not represent a continuing decline in water quality, let alone worsening environmental depredation nationwide. Contrasted with the relevant indifference to burning rivers in decades past, the public outcry over the 1969 fire signified that increasingly wealthy Americans now wanted to devote greater resources to environmental protection – and they would likely have even in the absence of federal regulations.

Source: Adler, J.H. (2004) "Smoking out the Cuyahoga fire fable: smoke and mirrors surrounding Cleveland", *National Review*, June 22.

Water in developing cities

There are several trends common to many urban areas of the developing world. First is the water scarcity that comes with continued population growth. Second, many cities are noting record volumes of water pollution from both household and industrial sources. Water pollution problems, such as pathogenic organisms in water, continue to pose immediate threats to human health. In addition to the issues of both quantity and quality, another problem that may arise in many developing cities is the issue of equity, an issue linked closely with poverty and economic development.

Continued urban growth means tremendous stress on already inadequate resources, while also exacerbating the social and economic challenges to infrastructure investment and pollution control. Many cities in the developing world have much more serious nonpoint sources of water pollution than cities in the developed world because large sections of their population are not served by sewers, drains or solid-waste collection. The right to safe water and adequate sanitation remains a promise unfulfilled. In 2000, the World Bank estimated that about 380 million urban residents in developing cities still did not have access to sanitation and at least 170 million still lacked access to safe drinking water.[171] Unsafe drinking water continues to be responsible for more than 80 percent of diseases and 30 percent of deaths in the developing world.[172] And at any given time, close to half the people in the developing world (3 billion plus) are suffering from one or more of the main water-linked disease such as diarrhea, cholera, entric fevers, guinea worm and trachoma. Some experts argue that good-quality drinking water and proper wastewater treatment during the past 15 years has been progressively deteriorating, not improving.[173]

Mexico City: sinking

While many are familiar with Mexico City's infamous air pollution problem, an equally pressing issue centers on its water supply. Mexico City is built on top of a vast underground aquifer. Approximately 72 percent of Mexico City's water supply comes from this aquifer.[174] In general, aquifers can be an excellent source of clean water since many pollutants are filtered out as the water passes through the soil and rocks. However, aquifers are slow to replenish. Water is now being extracted twice as fast from the aquifer than it is being replaced. As a result, land subsidence has become a serious problem for Mexico City. It is estimated that over the past century, Mexico City has sunk by as much as 10 meters in some areas.[175] Since the 1970s, the rate of sinking in the central part of the city is about 6 centimeters each year. Some of the areas on the outskirts of the city are sinking the fastest; many of these are among the poorest areas in the city.

Land subsidence weakens foundations to buildings and other infrastructure including damaging the water and sewage pipes. Pipes are ruptured or develop small leaks. It is estimated that 30 percent of water is lost in the pipes from leaks before reaching users – enough to provide water to more than 4 million people.[176] Leaking sewage pipes also contaminate the groundwater with heavy metals and micro-organisms.

Land subsidence has also caused flooding in the valley particularly after heavy rains. Originally, Texcoco Lake was three meters (9 feet) lower than central Mexico City; today the lake is over two meters (6.5 feet) higher than the city.[177] Dikes had to be built to confine the stormwater flow and pumps are needed to lift the drainage water under the city to the level of the drainage canals. In addition, because of subsidence, gravity no longer takes sewage and runoffs to the Grand Canal; the city has had to install pumps in order to remove sewage to Texcoco Lake. The cost of pumping water into, within and out of the Basin of Mexico amounts to almost $900,000 per day.[178]

Because of irregular water supply, many households use water storage tanks, located on rooftops. These are often left uncovered and are not always cleaned regularly, which enables bacteria to flourish.

Mexico City also has a significant part of the population with no access to sewage facilities. Some 30 percent of Mexico City residents do not have toilets, and as much as 93 percent of waste water is discharged untreated as it flows out of the city.[179] About 75 percent of the city's residents have access to the city's current wastewater system of unlined sewer canals, sewers, rivers, reservoirs, lagoons, pumping stations and deep drainage systems. During the rainy season, when domestic wastewater and industrial wastewater are mixed with stormwater runoff, more than 1.5 million tons pass through the city's sewage system untreated.[180] A plan to build four treatments plants has floundered for lack of financing and no serious proposals for the construction of new treatment plants have been made since the 1990s. As a result, the city lacks the capacity to reuse significant volumes of water for industrial and agricultural production within the valley, a practice that would also alleviate the demand for freshwater.[181]

Mexico City, like many developing cities, is challenged by problems related to water management that may cause social conflicts. The wealthier population of Mexico City consumes up to 40 times more water than that used by the poorer sector. Compounding the issue is that many of the poorest areas of the city not only receive water of unacceptable quality, they also suffer the inconvenience of rotating schedules for water deliveries, whether through the water network or from tanker trucks.[182]

The Ganges: a sacred river

The Ganges winds 1,500 miles across Northern India, from the Himalaya Mountains to the Indian Ocean, and through 29 cities with populations over 100,000. Known as Ganga Ma (Mother Ganges), the river is revered as a goddess whose purity cleanses the sins of the faithful and helps the dead on their path toward heaven. Hindus believe that if the ashes of their dead are deposited in the river, they will be ensured a smooth transition to the next life or freed from the cycle of death and rebirth. It is said that a single drop of Ganges water can cleanse a lifetime of sins. In cities along the Ganges, daily dips are an important ritual among the faithful. Many cities are pilgrimage sites.

Despite its spiritual importance, the physical purity of the river has deteriorated dramatically. While industrial pollutants account for some of the river's pollution, the majority of Ganges' pollution is organic waste: sewage, trash, food, human wastes and animal remains. Today nearly half a billion people live in the basin of the Ganges and more than 100 cities dump their raw sewage directly into the river. Pollutants have been measured at 340,000 times permissible levels.

Not surprisingly, waterborne illnesses are common killers, accounting for the deaths of some 2 million Indian children each year.

In the city of Varanasi, one of India's oldest and holiest cities, some 40,000 cremations are performed each year; those unable to afford the cost of traditional funerals often dump the body into the river. In addition, the carcasses of thousands of dead cattle, which are considered sacred to Hindus, are tossed into the river each year. The city also pumps some 80 million gallons of sewage waste daily into the river. And yet thousands of Hindus continue to come to Varanasi and the Ganges.

The Ganges presents an interesting case study. Because the river is holy, it attracts tens of thousands of pilgrims each day for ritual bathing, exposing large numbers of people to untreated, contaminated water. As one man commented, "There is a struggle and turmoil inside my heart. I wanted to take a holy dip. I need it to live. The day does not begin for me without the holy dip. But, at the same time, I know what is B.O.D and I know what is fecal coliform."[183] It is estimated that 40 percent of the people who take a dip in the river regularly have skin or stomach ailments.

In 1985, the Indian government initiated the Ganga Action Plan, to clean up the river in selected areas by installing sewage treatment plants. Twenty years later, the plan has been relatively unsuccessful. Some have blamed the failure on the adoption of expensive multi-million dollar Western-style treatment plants. The government spent $335 million. But many of the treatment plants malfunctioned, were designed improperly or could not handle the bacterial load. In addition corruption and ineffective monitoring contributed to the problems. There is growing criticism over the adoption of Western-style technology to solve developing world issues. Western-style technology tends to be very expensive, relies on highly trained engineers and workers to maintain the technology, and requires a stable and consistent supply of electricity. In addition, Western-style waste treatment plants were engineered for use in countries where there are no monsoon rains, and where the population does not drink directly from the water source. Few considered the radically different ways that people use rivers in India. Another criticism is that many local communities along the Ganges were not included in the planning process and therefore failed to participate in the recovery of the river.

As an alternative to the high-technology treatment plants, Veer Bhadra Mishra, a Hindu priest and civil engineer, collaborated with University of California Berkeley engineer, William Oswald, to develop a non-mechanized, low-tech sewage treatment plan. This plan is more compatible with the climate of India and replaces the high-tech solution with a wastewater oxidation pond system that would store sewage in a series of ponds and use bacteria and algae to break down waste and purify the water. The ponds allow waste to decompose naturally in water. Bacteria grow on the sewage and decompose it; the algae feed on the nutrients released by the bacteria and produce oxygen for the water. This alternative treatment does not require electricity, but relies on sunshine to speed up the decomposition. The pond system is much less expensive than mechanical treatment plants.

The debate between Western-style technology and lower-cost alternatives in the case of the Ganges highlights general themes applicable to many of the solutions to pollution in developing cities. While the 1980s and 1990s saw large-scale investments by governments in high-technology solutions, the lack of results has caused many to seek alternatives that are locally sensitive and economically affordable. More recently, experts in both the developing world and developed north contend that solutions should respond to local demands and should be as

Figure 8.3 *Beijing's Bamboo Park. Many cities have developed or maintained important water features, such as this water garden in downtown Beijing.* **Source: Photo by Michele A. Judd**

simple, sturdy and inexpensive as possible. Low-cost, low-technology solutions such as the pond system or pour-flush latrines, or even improved pit latrines, have been successful. An important element in pollution control is the role of public participation. The involvement of the local community and households is now seen as a crucial component to success.

Water issues in developing cities are not all problematic. Some cities are developing and protecting vital water resources. Figure 8.3 shows the river gardens of Beijing. The river gardens attract tourists and at the same time provide an important ecosystem within the rapidly growing city.

Reclaiming Toronto's Don River

Michael Hough recounts how the city of Toronto developed by exploiting the Don River and then turned its back on this resource. Recently residents and local government have been involved in reclaiming the river for public use and ecological restoration.

Toronto was first settled in 1787 ... and settlers harnessed the river's energy, built mills for lumber, flour, wool and paper, and mined the valley's clay and shale for brick-making, from which much of the early city was built. In less than 150 years, they cleared the lower valley of merchantable trees. The Don River was also perceived as a threat and an obstacle. Floods swept away mills and bridges, the river was an obstacle to the eastward expansion of the city and the great wetland, its mouth reviled as unhealthy swamp, lent credibility to the argument that straightening out the river and filling in the marshes would "secure the sanitary condition ... to the said river". By the end of the century, engineers had turned the last 5 kilometers of the river's meanders, where it dropped its sediments, into a canal. The railways were built in the valley, and the Ashbridges Bay marshes were filled in to create the port lands, the most massive engineering project on the continent in its time, forcing the Don into a right-angle turn into the harbour. By the mid-twentieth century, the city had turned its back on the river, a gap between places rather than a place in itself. As a sensory experience it had become a forgotten place; unloved and unused.

Moves to restore the river became an act of faith by the citizens of Toronto that grew out of the concerns of many people for the natural heritage of their city. Beginning as an informal citizen's organization, the "Task Force to Bring Back the Don" was formalized and supported by Toronto City Council in 1990. Its purpose

Continued

was to begin the process of renewal of the most degraded part of the river that flows through the city of Toronto, and ultimately to imitate the restoration of the entire watershed....

As an ongoing process of renewal and healing, the Don strategy involves key principles, including a fundamental understanding of process as a biological idea that is also integrated with social, economic and political agendas, economy of means where the most benefits are available for minimum input in energy and effort, and environmental education, where the understanding of nature in cities becomes part of a learning experience that begins with community empowerment and action.

Source: Hough, Michael (1995, 2nd edn) *Cities and Natural Processes: a Basis for Sustainability*, New York: Routledge. Quotes from pp. 39, 42, 54.

Conclusions

Since the 1970s many countries and municipalities have enacted pollution reform measures. Some cities have seen dramatic decreases in certain types of pollutants, while other pollutants have increased. Some cities have successfully implemented pollution laws and regulations, while many cities (and countries) lack the resources to adequately enforce such measures. The twin pressures of population growth and increasing consumption of resources have, in some cases, offset new laws and regulations designed to decrease or prevent pollution.

Guide to Further Reading

Adler, R., Landman, J. and Cameron, D. (1993) *The Clean Water Act: 20 years Later*, Washington, DC: Island Press.

Gumprecht, B. (2001) *The Los Angeles River: Its Life, Death and Possible Rebirth*, Baltimore: The Johns Hopkins University Press.

Lewin, T. (2003) *Sacred River: The Ganges of India*, Boston, MA: Houghton Mifflin/Clarion Books.

Melosi, M.V. (2001). *Effluent America: Cities, Industry, Energy, and the Environment*, Pittsburgh, PA: University of Pittsburgh Press.

Swyngedouw, E. (2004) *Social Power and the Urbanization of Water: Flows of Power*, Oxford: Oxford University Press.

Uitto, J. and Biswas, A. (eds) (2000) *Water for Urban Areas: Challenges and Perspectives*, Tokyo and New York: United Nations University Press.

White, R. (1994) *Urban-Environmental Management: Environmental Change and Urban Design,* New York: John Wiley.

For a good account of environmental policymaking, see

Desfor, G. and Roger, K. (2004) *Nature and the City: Making Environmental Policy in Toronto and Los Angeles*, Tucson: Arizona University Press.

9 Air pollution and the city

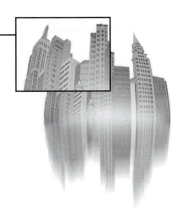

The harmful effects of air pollution have emerged as a major issue that cities in both the developed and developing world must address. In this chapter we will draw upon the experience of the US, as an exemplar of a rich country, as well as a range of developing countries.

The primary air pollutants

Air pollutants are composed of either visible particulates (ash, smoke, or dust) or invisible gases and vapors (fumes, mists and odors). The US EPA has identified more than 188 air pollutants; however regulatory measures have been primarily directed at only a dozen or so. These dozen air pollutants account for a majority of the volume of air pollutants. Most of the primary air pollutants are generated when fossil fuels are used. Throughout the twentieth century developed and developing societies have increasingly relied on three primary fossil fuels – coal, natural gas and petroleum – for energy and transportation. Because all fossil fuels are carbon-based, they release carbon monoxide and carbon dioxide when combusted. Coal also contains sulfur oxides and nitrogen oxides that are released during combustion. New sources of pollution, combined with increased use of fossil fuels for a myriad of needs has meant that air pollution for many cities has continued to increase despite regulatory efforts.

Nearly all of the so-called "most common" or most prevalent pollutants are generated in urban areas. In the US, the EPA has classified as six "criteria pollutants" those which pose the most ubiquitous threat to health and the environment (see Box Insert). These pollutants are called criteria pollutants because the EPA uses them as the basis for setting permissible levels. The EPA has developed one set of limits (called primary standard) to protect health; another set of limits called secondary standard are intended to prevent environmental and property damage. Almost all air quality monitors are located in urban areas, so air quality trends are more likely to track changes in urban emissions than changes in total national emissions.

The common air pollutants

Ozone (ground-level ozone is the principal component of smog)

- **Source** – chemical reaction of pollutants; VOCs and NOx
- **Health Effects** – breathing problems, reduced lung function, asthma, irritates eyes, stuffy nose, reduced resistance to colds and other infections, may speed up aging of lung tissue
- **Environmental Effects** – ozone can damage plants and trees; smog can cause reduced visibility
- **Property Damage** – Damages rubber, fabrics, etc.

VOCs* (volatile organic compounds); smog-formers

- **Source** – VOCs are released from burning fuel (gasoline, oil, wood, coal, natural gas, etc.), solvents, paints, glues and other products used at work or at home. Cars are an important source of VOCs. VOCs include chemicals such as benzene, toluene, methylene chloride and methyl chloroform
- **Health Effects** – In addition to ozone (smog) effects, many VOCs can cause serious health problems such as cancer and other effects
- **Environmental Effects** – In addition to ozone (smog) effects, some VOCs such as formaldehyde and ethylene may harm plants

All VOCs contain carbon (C), the basic chemical element found in living beings. Carbon-containing chemicals are called organic. Volatile chemicals escape into the air easily. Many VOCs, such as the chemicals listed in the table, are also hazardous air pollutants, which can cause very serious illnesses. EPA does not list VOCs as criteria air pollutants, but they are included in this list of pollutants because efforts to control smog target VOCs for reduction.

Nitrogen Dioxide (One of the NOx); smog-forming chemical

- **Source** – burning of gasoline, natural gas, coal, oil, etc. Cars are an important source of NO_2
- **Health Effects** – lung damage, illnesses of breathing passages and lungs (respiratory system)

Continued

- **Environmental Effects** – nitrogen dioxide is an ingredient of acid rain (acid aerosols), which can damage trees and lakes. Acid aerosols can reduce visibility.
- **Property Damage** – acid aerosols can eat away stone used on buildings, statues, monuments, etc.

Carbon Monoxide (CO)

- **Source** – burning of gasoline, natural gas, coal, oil, etc.
- **Health Effects** – reduces ability of blood to bring oxygen to body cells and tissues; cells and tissues need oxygen to work. Carbon monoxide may be particularly hazardous to people who have heart or circulatory (blood vessel) problems and people who have damaged lungs or breathing passages

Particulate Matter (PM-10); (dust, smoke, soot)

- **Source** – burning of wood, diesel and other fuels; industrial plants; agriculture (plowing, burning off fields); unpaved roads
- **Health Effects** – nose and throat irritation, lung damage, bronchitis, early death
- **Environmental Effects** – particulates are the main source of haze that reduces visibility
- **Property Damage** – ashes, soot, smoke and dust can dirty and discolor structures and other property, including clothes and furniture

Sulfur Dioxide

- **Source** – burning of coal and oil, especially high-sulfur coal from the Eastern United States; industrial processes (paper, metals)
- **Health Effects** – breathing problems, may cause permanent damage to lungs
- **Environmental Effects** – SO_2 is an ingredient in acid rain (acid aerosols), which can damage trees and lakes. Acid aerosols can also reduce visibility
- **Property Damage** – acid aerosols can eat away stone used in buildings, statues, monuments, etc.

Lead

- **Source** – leaded gasoline (being phased out), paint (houses, cars), smelters (metal refineries); manufacture of lead storage batteries

Continued

- **Health Effects** – brain and other nervous system damage; children are at special risk. Some lead-containing chemicals cause cancer in animals. Lead causes digestive and other health problems.
- **Environmental Effects** – lead can harm wildlife

Source: US Environmental Protection Agency; http://www.epa.gov/air/oaqps/peg_caa/pegcaa11.html, accessed August 2006.

Criteria pollutants are the most regulated air pollutants and policy has focused on limiting (but not outlawing) their use and discharge into the environment. One of the most difficult aspects to limiting the production of criteria pollutants is that they result from a variety of sources – the generation of energy, the use and combustion of fossil fuels in various forms of transportation and agricultural activities. For many cities, however, at least half of the air pollution comes from motor vehicle exhaust, a problem that has increased due to more vehicles on the road and more miles driven each year (see Figures 9.1 and 9.2).

In the US, a city that does not meet the primary standard for a given pollutant is called a non-attainment area. Cities in non-attainment areas must create an action plan for each pollutant in non-attainment. Action plans include targets for reduction of pollutants, ways to encourage use of public transportation and decrease the use of single-occupancy vehicles, securing voluntary reductions, and using a variety of outreach and educational tools. Although the EPA has been regulating criteria pollutants since the 1970 Clean Air Act was passed,

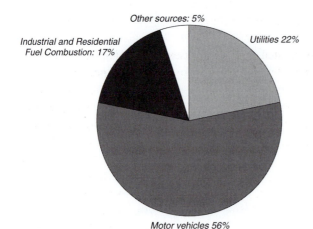

Other sources: 5%

Industrial and Residential Fuel Combustion: 17%

Utilities 22%

Motor vehicles 56%

Figure 9.1 *Major sources of nitrogen oxides. Source: Adapted form the US Enviornmental Protection Agency*

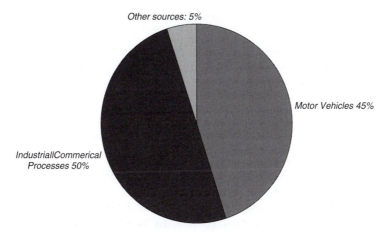

Figure 9.2 *Major sources of volatile oganic compounds*
Source: Adapted from the US Environmental Protecton Agency

many urban areas are still classified as non-attainment for at least one criteria pollutant. Approximately 120 million Americans (more than one third of the population) live in urban areas where air quality fails to reach federal standards for all of the six criteria pollutants.[184] Of the many air pollutants, however, ozone smog represents the single most challenging pollution problem for most US cities. More than 81 million Americans live in urban areas that exceed air quality concentrations for ozone, and progress on ozone has been the slowest for all criteria pollutants. For many cities, ozone levels have actually worsened over the past 10 years.

Photochemical smog

Smog occurs when volatile organic compounds (VOCs) react with nitrogen oxides and oxygen in the presence of heat and sunlight.[185] Pollutants undergo reactions that form ground-level ozone, or smog. Hence smog is not emitted directly into the air, but is created through a series of chemical reactions. It is produced by the combination of pollutants from many sources including smokestacks, cars, paints and solvents. Smog is found primarily in urban areas and is often worse in the summer months when heat and sunshine are more plentiful. Table 9.1 lists the worst US cities for smog in 2005. Smog is highly corrosive to rubber, metals and lung tissue. Short-term exposure can cause eye irritation, wheezing, coughing, headaches, chest pain and shortness of breath. Long-term exposure scars the lungs, making them less elastic and efficient, often worsening asthma and increasing respiratory tract infections. Because ozone penetrates deeply into the respiratory system, many urban residents are at risk including the weak and elderly, but also those who engage in strenuous activity. For those who

Table 9.1 *20 most ozone–polluted cities, 2005*

2005 Rank	Metropolitan Area
1	Los Angeles–Long Beach–Riverside, CA
2	Bakersfield, CA
3	Fresno–Madera, CA
4	Visalia–Porterville, CA
5	Merced, CA
6	Houston–Baytown–Huntsville, TX
7	Sacramento–Arden–Arcade–Truckee, CA–NV
8	Dallas–Fort Worth, TX
9	New York–Newark–Bridgeport, NY–NJ–CT–PA
10	Philadelphia–Camden–Vineland, PA–NJ–DE–MD
11	Washington–Baltimore–Northern Virginia, DC–MD–VA–WV
12	Charlotte–Gastonia–Salisbury, NC–SC
13	Hanford–Corcoran,CA
13	Cleveland–Akron–Elyria,OH
15	Knoxville–Sevierville–La Follette,TN
15	Modesto,CA
17	Pittsburgh–New Castle, PA
18	Youngstown–Warren–East Liverpool, OH–PA
19	Columbus–Marion–Chillicothe, OH
20	Detroit–Warren–Flint, MI

Cities are ranked by using the highest weighted average for any county within that metropolitan area.
Source: American Lung Association, Annual Report 2005. Retrieved at http://lungaction.org/reports/
sota05_cities.html#ozonecities, March 2006

were born and have lived in smoggy cities such as Los Angeles, Houston and Washington, DC, long-term exposure may be breaking down their body's immune system, increasing their chances of suffering respiratory illness and harming their lungs in later life. In many US cities, doctors are reporting increasing frequency of asthma in children.

Photochemical smog is one pollutant that is often exacerbated due to geography. Cities located in basins and valleys – such as Los Angeles, Mexico City and Denver – are particularly susceptible to the production of smog. Because of Denver's high altitude the city experiences frequent temperature inversions when warm air is trapped under cold air and cannot rise to disperse the pollutants to a wider area. As a result, smog may hover in place for days at time, generating a "smog soup" that envelops the city. Cities with high concentrations of fossil-fuel-burning power plants, metal smelters, cement and fertilizer factories, and with

Table 9.2 *The US EPA's Air Quality Index*

Air Quality Index (AQI)	Numerical value	Meaning	Color
Good	0 – 50	Air quality is considered satisfactory, and air pollution poses little or no risk.	Green
Moderate	51 – 100	Air quality is acceptable; however, for some pollutants there may be a moderate health concern for a very small number of people who are unusually sensitive to air pollution.	Yellow
Unhealthy for sensitive groups	101 – 150	Members of sensitive groups may experience health effects. The general public is not likely to be affected.	Orange
Unhealthy	151 – 200	Everyone may begin to experience health effects; members of sensitive groups may experience more serious health effects.	Red
Very unhealthy	201 – 300	Health alert: everyone may experience more serious health effects.	Purple
Hazardous	> 300	Health warnings of emergency conditions. The entire population is more likely to be affected.	Maroon

Source: www.airnow.gov/index.cfm?action=static.aqi retrieved on August 31, 2006.

high densities of cars and trucks, are also likely to experience smog production. The EPA estimates that almost 60 percent of smog is produced by transportation sources – cars, trucks and trains.

The EPA developed an "Air Quality Index" to educate people about how to respond to high levels of smog (Table 9.2). When ozone concentrations are high, smog alerts are issued to warn people with asthma and those with chronic respiratory disease to stay indoors and for healthy people not to exercise. For many urban residents, smog forecasts are part of their daily planning. On the worst smog concentration days in Washington, DC, officials encourage the use of public transportation by making the Metro and Metro buses free to all users. Major newspapers now feature the "Air Quality Index" for most cities, noting which air pollutants are likely to be above federal standards.

Table 9.3 *Toxic air pollutants*

Mercury
Lead
Polychlorinated Biphenyls (PCB)
Dioxins
Benzene
Pesticides such as DDT
Cadmium compounds
Chloroform
Formaldehyde
Methyl chloride
Arsenic

Source: US Environmental Protection Agency

Another category of air pollutants include those that are either toxic or hazardous. Table 9.3 lists several toxic or hazardous air pollutants. Hazardous and toxic air pollutants can cause cancer or can kill swiftly. In 1984 in Bhopal, India, leakage from the Union Carbide pesticide manufacturing plant created a lethal cloud of methyl iscoyanate gas which crept across a densely populated area. The toxic release killed approximately 4,000 people and injured more than 200,000. The death and injury toll continues to rise as lung diseases worsen and reports of mouth cancers, kidney and liver damage have been prevalent. The Bhopal tragedy inspired the 1990 Clean Air Act requirement in the US that factories and other businesses develop plans to prevent accidental releases of highly toxic chemicals.

Currently the EPA has identified 188 toxic or hazardous chemicals, many of them not well researched and poorly understood in terms of their impact on human health and the environment. In urban areas, toxic air pollutants are of special concern because of the concentration of people close to sources of emissions. In response to growing concerns about the long-term exposure to toxic or hazardous chemicals, the EPA has identified 33 of the 188 toxic air pollutants as the greatest threat to public health in cities. The EPA's strategy is to target these pollutants in cities by using a variety of national and local controls. Policy approaches to toxic or hazardous pollutants differ from criteria pollutants in two important ways. First, while states were charged to develop and implement their own plans to meet criteria pollutants, there is a federal plan to deal with toxics. Second, EPA policy aims to eventually outlaw or completely eliminate the discharge of toxic pollutants.

Similar to water pollution, the most significant air pollution problems today stem not from point sources, but from nonpoint sources. While much of the earlier legislation focused on point sources (such as factories), regulation must deal with increased nonpoint pollutants. Common nonpoint sources of air pollution include automobiles, trucks, buses, airplanes, lawn and garden equipment and even charcoal-burning backyard BBQs. The most common nonpoint sources are mobile. Although twenty-first century automobiles produce 60–80 percent less pollution than cars did in the 1960s, more people are driving more cars more miles. In 1970, Americans traveled 1 trillion miles in motor vehicles; by 2000 they drove 4 trillion miles.

The first US federal law dealing with national air quality standards was the 1970 Clean Air Act (CAA). It focused on point source air pollutants such as energy plants, factories and other stationary emitters. Since the 1970 CAA there have been several amendments, including the 1990 and 1997 Clean Air Act Amendments. The 1990 Amendments have a direct impact on cities. For the first time, the EPA established five categories for cities in non-attainment (those that did not meet federal regulations for criteria pollutants). The categories were marginal, moderate, serious, severe and extreme. Those cities originally classified as severe or extreme were mandated to become attainment areas by 2005. Los Angeles, labeled "extreme" in 1990 remains in non-attainment, although its notoriously smoggy air is the cleanest it has been in 50 years. Still, in one out of every three days, millions of Los Angelenos breathe in dirty air. In addition, the 1990 CAA established stricter tailpipe standards, finally addressing the growing problem of nonpoint pollution. Figure 9.3 shows a "smog station" where cars are required to be tested for tailpipe emissions every few years.

Pollution is a social construction. As new technologies allow researchers to measure and assess different pollutants at lower levels, and as medical and environmental science understands the impact of pollutants on public health and ecosystem integrity, standards of acceptability shift. As a social construction, pollution and regulation are subject to the political process. In the past, many of the EPA standards have been challenged in the federal courts. In 1998, several businesses and state groups challenged the 1997 CAA standards claiming the EPA misinterpreted the Act and gave itself unlimited discretion to set air standards. In 2001, the US Supreme Court unanimously upheld the constitutionality of the EPA's long-standing interpretation that it must set these standards based solely on public health considerations and not on the consideration of costs.

In the US, policy approach to urban pollution has been twofold. First, to impose regulatory measures that restrict and measure the quantities of pollutants

Figure 9.3 A smog station in Los Angeles. In order to maintain a car's legal registration, owners must periodically test the tailpipe emissions of their cars. Los Angeles is one city that imposes more stringent emissions standards than the US federal standards.
Source: Photo by John Rennie Short

released. Second, to establish economic measures that include pollution "taxes" or other financial disincentives. Yet after more than three decades after the first Clean Air Act, air pollution reform in US cities is as mixed as water pollution. Some pollutants – primarily those coming from point sources – have decreased. Between 1970 and 2000 sulfur dioxide decreased by 44 percent; carbon monoxide was down 25 percent, VOC decreased 43 percent, and with the mandatory phase-out of leaded gasoline and the introduction of unleaded gasoline, lead was down 98 percent.[186] Overall, total emissions of the six criteria pollutants decreased 29 percent. However parallel to water pollution, much of the improvements in air pollutants have been offset by increased populations that demand more energy as well as increased automobile use. Between 1970 and 2000, US population increased 36 percent, energy consumption increased 45 percent, vehicle miles traveled increased 143 percent and nitrogen oxides increased by 20 percent. Despite decades of regulation and good intention, cities in the developed world remain a long way from eliminating the threat of air pollution to both public health and environmental quality.

Cubatao and "The Valley of Death"

In the 1950s, government planners in Brazil designated Cubatao as the center of the nation's nascent oil refinery industry. It was an ideal location, in a valley beside a river, close to the coast that allowed the petroleum and steel industries to import raw materials and ship out finished products. Large state corporations like COSPIA (steel) and the oil monopoly Petrobras established a gigantic refinery and nearby chemical and fertilizer plants. Private corporations soon followed. The city became heavily concentrated with industrial manufacturing with few environmental controls. By 1985, Cubatao accounted for 3 percent of Brazil's GDP. Lack of enforcement resulted in thousands of tons of pollutants daily. By the early 1980s the city recorded the highest infant mortality rates in Brazil and over one third of the population suffered from pneumonia, tuberculosis or emphysema. Two alarming developments occurred in the late 1970s and early 1990s. First, dozens of babies were born without brains, although researchers were never able to prove that the birth defects were caused by pollution. One in three new-born babies failed to reach their first birthday. Second, the city experienced several mudslides down the mountains, which had been denuded. Cubatao became known as "the valley of death" and "the most polluted city on earth" and was even the subject of an ironic pop song called "Honeymoon in Cubatao". By 1983, the state government demanded industries start implementing pollution control; many responded, but others were notoriously lax. In the last fifteen years air quality has improved significantly, and respiratory ailments are half of what they were in 1984. Storage ponds that used to contain toxic waste have been cleaned out and turned into small lakes; the once barren hillsides have been reforested. And although Cubatao is still is one most polluted areas in the state, many of the pollutants are within World Health Organization recommendations.

The lessons of Cubatao highlight two important points. First, at the same time developed countries were instituting air pollution reforms, developing cities were experiencing rapid growth with little legislative or institutional measures in place. Second, although the global community recognized pollution as early as 1972 when the United Nations established its Environment Programme, pollution regulation in developing countries tended to occur a decade or two behind the reforms in the industrialized north.

Source: Anonymous (1999) "Face value: honeymooner in Cubatao", *The Economist*, July 10, 1999, pp. 68-71.

Air pollution in non-US cities

Table 9.4 lists three criteria pollutants for selected cities and Table 9.5 lists the cities that are most unsafe for children. The most polluted cities in the world include Beijing and Mexico City, which report high levels of all three criteria pollutants. Athens, Moscow and Mumbai have high levels of suspended particles; Cairo and Moscow have high levels of sulfur dioxide, and Amsterdam,

Table 9.4 *Three air pollutants in selected cities*

City	Suspended particles levels/ mg m³	Sulfur dioxide levels/ mg m³	Nitrogen dioxide levels/ mg m³
World Health Organization recommended max. levels*	90	50	50
Amsterdam	40	10	**58**
Athens	**178**	34	**64**
Beijing	**377**	**90**	122
Berlin	50	18	26
Brussels	78	20	48
Bombay	**240**	33	39
Cairo	–	**69**	–
Copenhagen	61	7	**54**
Kuala Lumpur	85	24	–
London	–	25	**77**
Los Angeles	–	9	**74**
Moscow	**100**	**109**	–
Mexico City	**279**	**74**	130
Milan	77	31	**248**
Montreal	34	10	42
New York	–	26	**79**
Singapore	–	20	30
Sydney	54	28	–
Tokyo	49	18	**68**

* numbers in bold exceed WHO levels.
Source: Adapted from A. Cohen et al. (2004) "Mortality impacts of urban air pollution", in *Comparative quantification of health risks: global and regional burden of disease attributable to selected major risk factors*, M. Ezzati et al., (eds), Geneva: World Health Organization, pp. 1353–1434.

Table 9.5 *Urban air pollution: most unsafe cities for children under five: combining measurements for sulfur dioxide, nitrogen dioxide and total suspended particulates (TSPs)*

1. Mexico City
2. Bejing
3. Shanghai
4. Tehran
5. Calcutta

Source: Adapted from Cohen A., et al. (2004) "Mortality impacts of urban air pollution", in *Comparative quantification of health risks: global and regional burden of disease attributable to selected major risk factors*, M. Ezzati et al., (eds) Geneva: World Health Organization pp. 1353–1434.

Athens, Copenhagen, London, Los Angeles, Milan, New York and Tokyo have high levels of nitrogen dioxide. It is not uncommon for cities such as Bangkok, Beijing, Calcutta, Delhi and Tehran to experience 30 to 100 days or more of poor air quality in a year. Jakarta exceeds health standards about 170 days per year; Mexico City exceeds acceptable levels of air pollution more than 330 days a year.

In some instances, national or municipal regulations are less stringent than in the US case. Or, as we have discussed previously, a lack of resources makes enforcement of laws difficult. Unlike many point-source polluters in the US or Europe, many developing countries have little data on industrial polluters.

The magnitude of toxic emissions is serious and growing in the developing world. In contrast to cities in the North, cities in the developing world face rapid increases in population, high density and shanty cities, dramatic increases in cars and trucks, less fuel efficiency and often limited governmental control. Although the highest number of vehicle owners (per thousand in a population) are found in the rich countries such as US, Australia, Italy, New Zealand and Canada, the fastest increase in vehicle ownership is in the developing world. Between 1980 and 1998 ownership in several developing countries increased 5-fold – in South Korea vehicles per thousand people increased 1,514 percent; in Thailand 692 percent, Nigeria 550 percent, China 300 percent and Pakistan 300 percent.[187] According to the World Health Organization, 340,000 people in cities die each year from diseases caused by indoor pollution, another 400,000 die from diseases associated with outdoor pollution.[188] Figure 9.4 shows deaths due to urban air pollution.

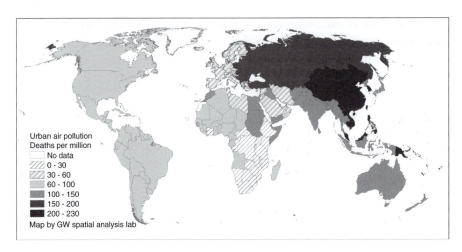

Figure 9.4 Deaths from uban air pollution (map)
Source: Lisa Benton-Short, adapted from World Health Organization data

Pollution in Indian cities

In India, most cities exceed the air quality standards set by the World Health Organization. In Delhi, 2,000 metric tons of air pollutants are emitted into the atmosphere every day. Vehicular sources account for 65 percent of air pollutants; industrial emissions account for 29 percent and domestic emissions 6 percent. Particulate matter (suspended particles such as smoke, soot and ash) is also a problem. In 2006 the World Bank ranked Delhi the worst city in the world for particulate matter (see Table 9.6). In fact Indian and Chinese cities were home to 18 of the smokiest 20 cities in the world. In cities such as Delhi the rapid rate of population growth combined with increased resource consumption compounds air pollution problems. Between 1970 and 1990 the number of vehicles in India increased from 1.9 million to 21 million. The majority of the vehicular population is found in urban centers, with about one-third of the 21 million vehicles concentrated in 23 metropolitan cities. Unlike the US, which mandates tighter emissions standards on both automobiles and two- and four-stroke engines (such as ATVs, lawn mowers, motorcycles), India does not. There is a high percentage of two-wheeled vehicles in India, and while these are generally more fuel efficient than passenger cars, there is little pollution control.

Air pollution is also damaging India's most famous monument, the Taj Mahal at Agra. Surrounding cities contain more than 2,000 polluting industries ranging

Table 9.6 *Five smokiest cities, particulate matter (micrograms per cubic meter)*

City	PM mg m3
1. Delhi	177
2. Cairo	159
3. Calcutta	145
4. Tianjin	139
5. Chongqing	137
World Urban Average:	60

Source: World Bank 2006, *World Development Indicators, 2006*, Washington, DC: World Bank Publications.

from brick kilns to an oil refinery as well as traffic that uses high-sulfur diesel fuel. Acidic emissions of sulfur dioxide and nitrogen dioxide are eroding and dissolving the marble monument.[189] In Mumbai, there are high levels of sulfur dioxide and there has been an increase in the prevalence of breathing difficulties, coughs and colds. According to the World Resources Institute, mortality data show a link between dense air pollution in Mumbai and a higher rate of death from respiratory and cardiac conditions.[190]

Autorickshaws in Delhi

In Delhi, there are an estimated 2 million vehicles, three quarters of which are two- or three-wheeled vehicles (called tempoes) (see Figure 9.5) Many of them are old and poorly maintained. As a result, the large number of two-wheelers contribute greatly to carbon monoxide, sulfur dioxide and nitrogen oxide emissions. In general, motor vehicles in developing countries tend to be less-fuel efficient and more polluting than those in the North because of a lack of access to new technologies, a greater proportion of older vehicles, poorly surfaced or badly maintained roads, weaker environmental legislation or weak enforcement of the regulations and the dominance of low-quality fuels such as diesel with a high sulfur content.[191]

An estimated 80 percent of Delhi's taxis and autorickshaws (powered by four-stroke or two-stroke engines) are older than 15 years; many are 30 or more years old. More than 40,000 autorickshaws are on the roads of Delhi and many of these are two-stroke engines that contribute significantly to emissions of

Continued

carbon monoxide and suspended particulate matter. Autorickshaws were identified as the major culprits in air pollution. In 1998, in response to pressure from environmentalists to check the growing pollution in almost every major city, the Indian government imposed stiff new emission norms and announced it would phase out all commercial vehicles older than 15 years, blaming the worsening air quality on the sharp rise in vehicles. Today, there are only 20,000 autorickshaws on the road and suspended particulate matter has decreased.

Rickshaws are also a main source of transportation in most Indian cities, as the photo of both traditional (foot-powered) and autorickshaws (motorized) in Hyderabad shows. (see Figure 9.5)

Rickshaws in Delhi

Figure 9.5 *Traditional and autorickshaws in Hyderabad, India.*
Source: photo by Michele A. Judd

Airborne lead

A particularly toxic component in urban air is lead, the heavy metal which for many years was added to gasoline to raise octane levels and help engines run more smoothly. Leaded gasoline is the primary source of lead exposure in many developing cities. Only recently have developing countries begun to phase out leaded gasoline. While it was phased out by the mid 1980s in many developed countries, in India unleaded gasoline has been sold only since 1998.[192] Although about 85 percent of gasoline sold in the world is unleaded, the 15 percent that is leaded is distributed primarily in the developing world. Lead levels in the air of large African cities such as Cairo, Cape Town and Lagos are up to ten times those of European cities. Lead is one of the most harmful airborne pollutants, causing neurological damages that particularly impact on children. High exposure to lead in children can have more dramatic health impacts than for adults; slowed growth, hearing problems, headaches, reduction in IQ and mental retardation are among the problems associated with lead poisoning. In 1998, the World Bank launched its Clean Air Initiative in Sub-Saharan Africa and identified the elimination of lead from gasoline as the priority, noting that "switching quickly to unleaded gasoline is one of the most cost-effective steps to protect children's health".[193]

Indoor air pollution

A distinct problem in many cities in the developing world is indoor air pollution that results from the use of wood or animal dung or kerosene for heat and cooking.[194] Indoor air pollution may pose a greater risk in developing cities than ambient air pollution. Over 2 billion people around the world use coal and biomass fuels such as wood, cow dung, charcoal and grass for cooking and heating. The pollutants that result from the burning of these fuels in inefficient stoves combined with poor ventilation have severe and sometimes fatal consequences. Particularly in urban squatter neighborhoods, women often perform their daily cooking in small, enclosed areas, relying on some form of biomass (see Figure 9.6). Often the open fires or poorly functioning stoves prevent complete combustion, which means the fuels do not burn cleanly, and emit many types of toxic gases such as carbon monoxide. Indoor air pollutants are 1,000 times more likely to reach people's lungs than outdoor air pollutants. Indoor air pollution disproportionately impacts on women and children, who devote a large portion of the day to cooking or other tasks inside the home.[195] Young children, who are often carried on their mother's back or kept close by, are also vulnerable. A World Health Organization report noted that indoor air pollution is the most lethal killer in the developing world after malnutrition, unsafe sex and lack of safe water.

Figure 9.6 Indoor air pollution. A woman cooks over an indoor fire while her children look on.
Source: Photo by Elizabeth Chacko

In India, 80 percent of households use bio fuels and the estimated child mortality rate from indoor air pollution is 500,000 children per year, mainly from acute respiratory disease. According to the World Resources Institute the exposure of pregnant women to indoor smoke results in a 50 percent increase in stillbirths.[196] And women using biomass for cooking are up to 75 times more likely to contract chronic lung disease.[197] Between 3.5 percent and 7 percent of the national burden of disease in developing countries is attributable to respiratory and other problems related to indoor air pollution, a much larger percentage than is ascribed to urban and industrial pollution.

Conclusions

During the twentieth century, air pollution, once a localized problem, became a global one. Nowhere is immune from toxic fallout and yet the most intense effects on both ecosystems and human health are local – and urban.

Guide to Further Reading

Brimblecombe, P. (ed.) (2003) *The Effects of Air Pollution on the Built Environment*, Singapore and River Edge, NJ: Imperial College Press; distributed by World Scientific.

Bulkeley, H. and Bestill, M. (2005) *Cities and Climate Change*, New York: Routledge.

Elsom, D. (1996) *Smog Alert: Managing Urban Air Quality*, London and Sterling, VA: Earthscan.

Molina, L. and Molina, M. (2002) *Air Quality in the Mexico Megacity*, Boston: Kluwer Academic.

Rock, M.T. (2002) *Pollution Control in East Asia: Lessons from the Newly Industrializing Economies*, Washington, DC: Resources for the Future.

Harpham, T. and Tanner, M. (eds). (1995) *Urban Health in Developing Countries: Progress and Prospects*, New York: St Martin's Press.

White, R. (1994) *Urban-Environmental Management: Environmental Change and Urban Design*, New York: John Wiley.

World Health Organization (2002) *The World Health Report 2002: Reducing Risks, Promoting Healthy Life*, Geneva: World Health Organization.

For up to date information on ozone/smog and interactive maps, go to: www.epa.gov/air/ozonepollution

Also www.scorecard.org allows you to profile the air, land and water pollution in your state, city or community in the US.

10 Garbage and the city

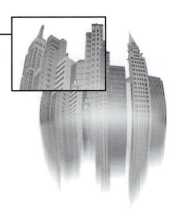

Garbage is a fundamentally urban problem. In the developed world, the twentieth century brought two major changes to the challenge of dealing with garbage in the cities. The first was the acceleration in the consumption of goods and resources per person which increased the generation of solid waste. In the decades after the Second World War, economic prosperity led to a consumer-oriented sector that offered more and more items. Second, new technologies, particularly those developed in the chemical industry such as plastics, radically changed the composition of garbage, presenting new challenges and problems for the environment. The problems of increased waste quantities were compounded by the alterations to the waste stream.

In this chapter, we explore how cities are confronting the growing waste problem. There are several issues associated with refuse. There is the technical problem of both collection and disposal and the environmental hazards associated with refuse. Underlying social and cultural values have made reform in refuse lag far behind water and air pollution reform. In the developed world, cities generate very high rates per capita of waste, and recycling rates are low and part of the formal economy. In addition, many cities have sought to "export" their garbage problems to developing countries. In the developing world, cities generate less per capita waste, but have fewer resources to ensure that collection and disposal will not create environmental and health hazards. Interestingly, in most developing cities, recycling rates are high because the urban poor separate recyclables as part of the informal economy.

In this chapter we also focus on US cities because America generates more refuse per person than any other country in the world. It produces some 30 percent of all the world's garbage (with only 5 percent of the world's population). Therefore, the garbage problem in US cities is more acute than anywhere else.

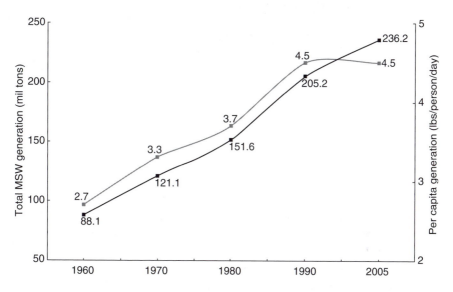

Trends in municipal solid waste Waste Generation 1960–2005

Figure 10.1 *Trends in MSW waste generation 1960–2005*
Source: Lisa Benton-Short using data from US EPA

In the US, postwar economic prosperity created a "throwaway" culture that generated vastly increased volumes of waste. New materials such as plastics, other synthetic products and toxic chemicals made their way to landfills. In addition, new changes in the packaging industry created innumerable goods with very short lives. Between 1955 and 1965 per capita refuse increased by 78 percent in New York City; between 1958 and 1968 it increased 51 percent in Los Angeles.[198] In 1969, the US collected 30 million tons of paper and paper products, 4 million tons of plastic, 100 million tires, 30 billion bottles, 60 billion cans and millions of tons of grass, tree trimmings, good waste and sewage sludge.[199] By the 1960s solid waste had become a critical environmental issue and land pollution joined air and water pollution as a triad of blights deserving federal attention.[200] Figure 10.1 shows how trends in municipal solid waste (MSW) increased from 1960–2005.

Technical issues

Solid waste is defined as "material that has no apparent, obvious or significant economic or beneficial value to humans that is intentionally thrown away for disposal". Unlike air and water pollution, solid waste must be collected to be

Table 10.1 *Sources and types of municipal solid wastes*

Source	Typical waste generators	Types of solid waste
Residential	Single and multi-family dwellings	Food wastes, paper, plastics, textiles, leather, yard wastes, glass, metals, ashes, bulky items like refrigerators, tires, hazardous wastes
Industrial	Manufacturing, construction sites, power and chemical plants	Packaging, food wastes, construction materials, hazardous wastes, ashes, slag, scrap materials, tailings, special wastes
Commercial	Stores, hotels, restaurants, markets, office buildings	Paper, cardboard, plastics, wood, food wastes, glass, metals, special wastes
Institutional	Schools, hospitals, prisons	Same as commercial plus medical wastes
Construction and demolition	New construction sites, road repair, renovation sites, demolition of building	Wood, steel, concrete, dirt
Municipal services	Street cleaning, landscaping, parks, beaches, wastewater treatment plants	Sludge from waste treatment plants, landscape and tree trimmings, general wastes from parks

Source: Urban Development Unit, World Bank, May 1999. "What a waste: solid waste management in Asia", Washington, DC: World Bank, p. 5.

disposed of. The classification of Municipal Solid Waste (MSW) includes waste generated from residential, commercial, industrial, institutional and construction/demolition activities and municipal services (Table 10.1 lists the sources and types of solid wastes common in many urban areas). In many cities MSW refers only to household or residential waste. If construction and demolition waste is included, the quantity of waste is often doubled or even tripled.

By the 1950s most US cities had instituted service charges for the collection and disposal of household refuse. The most popular form of disposal was land disposal. For the first half of the twentieth century, most refuse was simply placed in "open" dumps. Open dumps were natural or human-made depressions in the landscape that served as places to dispose of refuse. The hazards of open dumps included fires and the leaching of numerous chemicals out of refuse into the soil or groundwater. They also posed health and safety issues. Human fecal matter is commonly found in municipal waste. Insect and rodents attracted to the waste can spread diseases such as cholera and dengue fever. The EPA has identified 22 human diseases that are linked to improper solid waste management.[201] Until very recently, open dumps were not regulated.

In the postwar years, advances in sanitary engineering, notably by the Army Corps of Engineers, improved upon the open dump and established standards for the proper use of the sanitary landfill. Contemporary landfills are huge engineering feats that include layers of non-permeable clay or synthetic lining and a network of pipes to collect leachate and methane as an energy source. Landfills were sited according to drainage patterns, wind and distance from the city, rainfall, soil types and the depth of the water table. Landfills, unlike open dumps, produce reclaimed land, places that can be used for parks and recreational areas, parking lots and industrial purposes. By the 1960s most cities had outlawed the open dump, by then considered a danger and a health menace, and looked to landfill as the solution to the problem of refuse.

Additional technological developments of the late twentieth century aided in both collection and disposal. The introduction of compaction vehicles, which compressed wastes by 30 percent, allowed more volume to be collected curbside. Similar compaction systems can be found in private residences. In-home garbage disposals that grind food wastes were once promoted as a technology that would completely eliminate household garbage; in reality disposals just transfer the pollution problem from curbside collection to waste water treatment plants. The development and use of transfer stations – collection points where trucks unload into larger vehicles or temporary storage facilities – increased efficiency and assisted in sorting.[202] Technological "fixes" often provide partial solutions, or in some cases alleviate the problem, but rarely do they solve the underlying issue of the production of the waste in the first place.

The garbage crisis?

Despite the improvements in refuse technology, solid waste remains a growing problem. Between 1970 and 1990, municipal solid waste increased 61 percent, with approximately 80 percent going to landfills. According to the EPA, in 2003 the US generated 236 million tons of trash (approximately 4.5 pounds per person) of which only 68 million tons were recycled.[203] In the past thirty years, many cities have experienced a "garbage crisis". Many landfills have closed, or will soon close, as they reach their capacity. In 1988 there were 7,924 landfills in the US. By 2000 there were 2,216, although the average landfill size did increase.[204]

In the US there is a geography to the garbage crisis. Cites in the Northeast, New England and Mid-Atlantic are running out of space to dump their refuse, while cities inland such as Dallas, Phoenix and Santa Fe have landfill capacity for more than 40 years. The garbage crisis is more regional than national. One result for those cities facing landfill closures has been much higher tipping fees for both collection and disposal. Ironically, the higher fees for refuse collection and disposal may provide an economic incentive for cities to improve their recycling programs. Cities that can divert more refuse to recycling will save money.

One of the more infamous landfills that has closed is Fresh Kills. Fresh Kills (originally from the Dutch "kill"=water) covers some 3,000 acres and is located on the western shore of Staten Island in New York. Beginning in 1948, Fresh Kills served the city of New York as its sole landfill until 2001. At its height, Fresh Kills received some 17,000 tons of trash per day (see Table 10.2 for a list of items sent to landfills in 2000). It was a putrid mountain of waste, the largest human-made structure in the history of the world. In 1996, officials announced that Fresh Kills would close and began phasing out the use of the landfill and stated the city would expand its recycling program and search for alternative landfill destinations. Today, trash from the Bronx, Brooklyn and Manhattan is

Table 10.2 *Material in New York City waste stream, 2000*

Material	Quantity (tons)	Material	Quantity (tons)
Furniture	217,790	Books	27,484
Appliances	64,067	Telephone directories	11,547
Carpets and rugs	671,011	Third-class mail	103,706
Clothing and footwear	114,324	Paper towels, paper plates and cups	108,170
Motor oil	2,580	Hangers from dry cleaners	2,269
Toothbrushes	592	Incandescent light bulbs	1,686
Disposable diapers	103,324	Pallets and wood containers	254,388
Milk cartons	15,080	Paper grocery bags	13,094
Clear HDPE jugs	11,604	Plastic grocery bags	4,210
Aerosol cans	8,935	Plastic plates and cups	12,760
Office paper	112,527	Single-use cameras	23.8
Newspapers	284,926	Toilets	1,346
Magazines	69,630	Writing instruments	1,033
Folder cartons	121,185	Trash bags	26,390
Others	112,323	TOTAL	2,478,004

Source: Adapted from New York City report, retrieved at www.ci.nyc.ny.us/html/dos/pdf/wprr/ wprr06.pdf, p. 11.

exported out of New York State. However, the exporting of trash cost the city some $622 million between 1998 and 2002 and there have been unpredicted costs as well. For those who live along city truck corridors, there has been a "remapping" of the flow of garbage. Floors and windows of homes and apartments vibrate when the trucks roll by; the smell of rotting garbage and truck exhaust fills the air outside. Proposed solutions to the refuse problem include transporting the trash by rail or barge to landfills in New Jersey, Virginia and states in the Midwest and South. Since its closure, the mounds at Fresh Kills have been covered with a foot of dirt, topped with a foot of sand, followed by a plastic liner and two more feet of soil. The plan is to "reclaim" the landfill as a "post modern forest". Landscape architects envision Fresh Kills as a thriving park and bird sanctuary, where in another 30 years or so people might walk on natural trails or picnic among the freshly made wilderness.

At the same time as landfills like Fresh Kills are closing, it has become more difficult to open new landfills as many environmental organizations have effectively opposed them on the basis of potential environmental hazards. Between 1990–2005 only a handful of new landfills were approved and opened. In the midst of increasing refuse and decreasing landfills, many cities have turned to one of two solutions. First, many cities now truck their garbage out of the urban region to other parts of the state and in some cases even across state lines. Philadelphia, for example, sends some of its trash to Delaware landfills. Ironically, the state of Pennsylvania imports garbage from cities in other states. Many cities in the Northeast transport their refuse to landfills in the Midwest, where large areas of land are still available. Second, and perhaps more ethically problematic, many cities have attempted to make arrangements to ship their municipal waste to developing countries. Poor countries are often attracted to the disposal fees cities offer. But critics call this practice "waste imperialism" and argue that governments are endangering their own citizens by accepting refuse and hazardous wastes from rich countries. With no solution to the garbage problem forthcoming, cities desperately search for new landfills and disposal sites.

Consider the very famous case of the city of Philadelphia and the cargo ship the *Khian Sea*. In September 1986, the *Khian Sea* left Philadelphia with 14 tons of ash from refuse that had been incinerated. Initially, Philadelphia had contracted with a firm that found a shipping company willing to take the ash to a manmade island in the Bahamas. Alerted by environmental organizations such as Greenpeace about the potential environmental hazard of the ash, the Bahamian government denied permission for the *Khian Sea* to dock. Over the next two years, the *Khian Sea* attempted to make arrangements to unload its ash with eleven different countries including Honduras, Haiti, the Dominican Republic, Senegal, Sri Lanka, Indonesia and the Philippines. Finally, in November 1998, the ship arrived in

Singapore – without its ash. Somewhere along the way, the ship had been sold, and twice renamed. The ash was most likely dumped illegally into the ocean.

Ethical questions of waste disposal in places far beyond the source of origin have become a recurring theme as air, water and now refuse crosses national boundaries. Is it equitable for cities to intentionally export their pollution problems to other, often poorer, areas, states or countries?

The growing concern over potential environmental hazards associated with refuse and incinerator ash may play out in the legal system. For example, some states in the US have passed laws that restrict or prohibit trash imports. In New Jersey, state and local leaders blocked a plan to ship New York City's trash to Newark landfills, and one utility company filed suit to block trash deliveries by New York City garbage trucks to a transfer station in Elizabeth, NJ. Said Elizabeth Mayor J. Christian Bollwage, "They [NY] are forcing their will on others because they don't have the creativity to solve their own problems."[205] In Docklands, UK, the siting of a solid waste transfer station generated local protest (see Figure 10.2).

Figure 10.2 *Protesting against a solid waste transfer station. In Docklands, UK, the construction of a solid waste transfer station is vandalized by graffiti that notes "class war", "locals first" and "we build for yuppies", a sign of neighborhood discontent and protest.*
Source: Photo by John Rennie Short

The environmental hazards of landfills

Unlike air and water pollution, garbage and solid waste has been regarded as more of an engineering problem than a public health issue.[206] However, there are several potential environmental problems and public health issues associated with landfills. A significant part of the composition of the US waste stream is biodegradable organic material such as food wastes, yard wastes and tree trimmings. In theory, over time organic material should biodegrade, creating nutrient rich soils. In reality, however, landfills are so compacted that there is not enough air for microbes and bacteria to break down organic debris. One of the greatest myths about the US waste stream is that it is primarily composed of plastics and disposable diapers. And while it is true that there is a significant amount of packaging waste, in reality, most of what people discard could be recycled, reused or composted but often is not (see Figure 10.3). Landfills, even the most recently engineered ones, still have a potential to contaminate soil and ground-water. Trash bags leak their contents. Water from rain also seeps through refuse, picking up a variety of liquids and poisons such as ammonia, chlorides, zinc, lead and acids (see Table 10.3 for a list of household hazardous wastes commonly sent to landfill). Leachate and poisons can then leak into aquifers, contaminating fresh water sources.

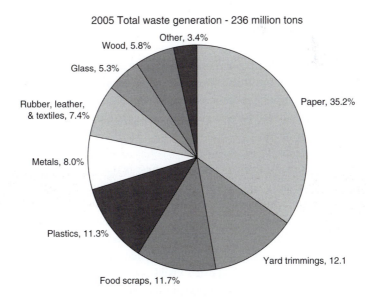

2005 Total waste generation - 236 million tons

Other, 3.4%
Wood, 5.8%
Glass, 5.3%
Rubber, leather, & textiles, 7.4%
Metals, 8.0%
Plastics, 11.3%
Food scraps, 11.7%
Yard trimmings, 12.1
Paper, 35.2%

Figure 10.3 *Composition of the US Waste Stream, 2005*
Source: Lisa Benton-Short using data from US EPA

Table 10.3 *Household hazards in landfills*

Used motor oil
Auto batteries
Antifreeze
Degreasers
Weed killers
Insect killers
Ant poisons
Bug sprays
Lighter fluid
Gasoline
Oil-based paint
Turpentine
Concrete cleaners
Drain and oven cleaners
Aerosol products
Cleaners with bleach or ammonia
Nail polish remover

Source: Adapted from US Environmental Protection Agency

A consuming mentality

Americans discard more garbage per capita than citizens of other prosperous nations, and far more than those in the developing world. In some regards, the production of garbage reveals much about levels of affluence, household formation, commercial activity and values. Our "garbage crisis" is as much social and political as physical.

Garbage, and particularly household refuse is one pollutant that is often invisible. We throw away our unwanted food scraps and household items into bags, then into bins, which are then wheeled to the sidewalk or curbside. A large collection truck comes by, usually when we are gone from the home or apartment, and it magically disappears. Garbage is effectively removed from our lives. Few people ever visit landfills and see the vast accumulation of trash. And unlike air and water pollution, which is often visible on a daily basis, garbage is something we rarely see or think of as impacting the environment. Writer Elizabeth Royte noted, "From the moment my trash left my house and entered the public domain

... it became terra incognita, forbidden fruit, a mystery that I lacked the talent or credentials to solve."[207]

Landfills can be considered silent "monuments" to our consuming lifestyle. Many landfills are visible from space – Fresh Kills landfill in Staten Island can be seen from orbiting satellites. And yet they are often invisible to their urban populations.

In many developed economies, but particularly in the US, we have favored convenience over conservation, short-term needs over long-range resourcefulness. The fast food of McDonalds is both indicative of a cultural value that embraces convenience and accepts the short-term duration of packaging. Food arrives within minutes. The food wrappers and drink cups have a life span of less than one hour, yet they survive in landfills for years. The quintessential fast-food society has become a disposable society. This extends beyond the fast food restaurants into many aspects of our culture. It is often less expensive to purchase a new radio or DVD player than repair a broken one. We now purchase purposely "disposable" products such as razors, toothbrushes, paper plates and cups, and writing pens.

The urban refuse problem stands for larger issues associated with the production of waste in general. As with water and air pollution, few legislative reforms punish or prohibit the production of waste; rather reforms often deal with financial incentives to develop new technologies to reduce the pollutants. Cities deal with refuse as an "end of the pipe" issue, rather than challenging the production of refuse. American consumers have few incentives to decrease their own refuse stream: there are virtually no economic costs associated with the volume of trash per household, generally these services are a flat fee. New technological developments in the plastics industry has made packaging thinner and lighter; however consumers still generate and then discard a synthetic product that does not biodegrade in the landfills. Table 10.4 lists numerous products that are synthetic plastic-based. In 1945 the US produced some 400,000 tons of plastic products. By 1998 the US generated 47 million tons, worth more than $200 billion.

As recycling programs strengthen, and manufacturers reduce packaging materials, solid waste generation in many cities in rich countries may become stable or even decrease. Already there are signs that governments are doing more by way of legislation to reduce materials that end up in landfills, and to make the manufacturing side of consumer goods more responsible for the wastes they generate. In 1991, Germany passed a law requiring industry to take back, reuse and or

Table 10.4 *Plastics in the waste stream*

....

Plexiglas (polymethylmethacrylate)
Polyesters
Polyvinyl Chloride (PVC)
Teflon (polytetrafluoroethylene)
Polyurethane
And other products such as:
Golf clubs, bike helmets, backpacks, "Fleece" sweaters and jackets, Gortex, toothpaste,
    toothbrushes, Chapstick, zippers, billiard balls, knobs and buttons.

....

recycle packaging materials. By making industry take back its packaging, the ordinance shifts the burden of managing packaging waste away from cities towards manufacturers, distributors and retailers.[208] The Netherlands and Sweden also have extended product responsibility frameworks. The Dutch government implemented a new policy that requires distribution of life cycle assessment at each stage for manufactured products; in Sweden, a new law promotes more efficient use of resources in the production, recovery and reuse of waste. Canada also passed a National Protocol on Packaging in 1990 in order to reduce the amount of packaging that goes to landfills. Many hope these types of reforms will mark the end of the throwaway society.

In the US the EPA has more recently emphasized the importance of reducing the amount of waste created and of recycling as much as possible. Some cities have been very successful at recycling and reusing. There are also creative ways to reduce food wastes reaching landfills. The San Francisco Food Bank collects some 37 tons of edible food a month from wholesalers and distributes it to local service agencies; composting food scraps and leftovers at the University of Massachusetts diverted 48 percent of the materials destined for landfills. In the US, there are more than 9,300 curbside recycling programs and cities are now averaging recycling of some 25–28 percent of the wastes.[209] Overall US cities have seen increased recycling rates (see Figure 10.4).

Incinerators: the solution to garbage?

In 1976 the US Congress passed the Resource Conservation and Recovery Act (RCRA). RCRA gave the EPA authority to regulate dangerous and hazardous materials, which expanded their ability to track both the generation and the

MSW Recycling rates 1960-2005

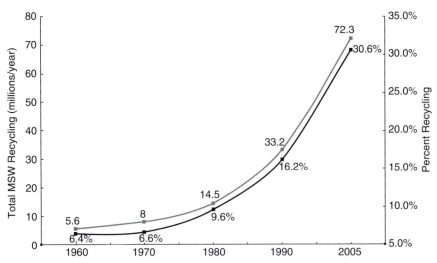

Figure 10.4 *MSW Recycling rates 1960-2005*
Source: Lisa Benton-Short using data from US EPA

disposal of these dangerous chemicals. The EPA defines a hazardous waste as one which meets one of the following characteristics: ignitability, corrosivity, reactivity and toxicity. The agency has since identified 450 wastes as hazardous. By the late 1990s the US produced some 197 million tons of hazardous waste annually.

Another very important aspect of RCRA was its implicit solution to the garbage crisis. RCRA was influenced by three growing concerns: increasing landfill closures; public anxieties about the public health and environmental impacts of landfills; and the growing dependence on foreign oil imports which made Congress look for other potential sources of energy. These three issues coalesced and influenced the RCRA legislation in a profound way. The law phased out open dumps and set higher standards for landfill design. More importantly, the law established tax incentives and government grants for cities to build waste-to-energy incinerators. In municipal America's war on trash, incinerators became the strategic weapon of choice. The incineration movement had seized the political moment.

Waste-to-energy incinerators burn garbage at very high temperatures, approximately 1,800° F, in a combustion chamber, thus superheating water in the boiler, which in turn spins a turbine that then generates electricity which can be sold to the power company. In theory, because incinerators burn at such high temperatures, the heat destroys any and all toxic material. In reality, early incinerators imported from Europe had problems coping with the high organic content of the US waste stream, often causing incomplete combustion and the release of toxic or hazardous chemicals. This solution to garbage reflects much of what we have discussed in previous chapters. Many cities, often with the encouragement of federal laws, seek the high technology solution to pollution. Incinerators cost as much as $300 million to construct and they are expensive to maintain as well. At the same time the new law did little to encourage waste reduction in the first place and did not support recycling and reuse programs with the same level of financial assistance and incentives that it did the inciner-ation option. Ironically, because incinerators operate most efficiently when at capacity, many private incinerator firms contracted with cities to receive as much refuse as possible; in some cases forestalling the development of munici-pal recycling programs.

Between 1976 and 1990 there were more than 140 incinerators constructed in cities around the US. Gradually concerns emerged about the potential environ-mental hazards. Nearly all incinerators produce bottom ash: ash that falls to the bottom of the combustion chamber. Bottom ash, which tends to be high in con-centrations of heavy metals, is disposed of by sending it to landfills. Air stack emissions (which in theory should emit only water vapor) have been found to contain heavy metals, sulfur dioxide, nitrogen oxides and carcinogens such as dioxins. Dioxins are generated by the incineration process. When pesticides and plastics are burned along with organic material, the result is the creation of diox-ins, which are fat soluble in the human body and highly carcinogenic. In Europe, where the incinerator technology was perfected, the waste stream is different from the US (the US has higher organic materials). As a result, many of the incinerators did not burn refuse efficiently due to the 'wetter' content of organic materials. The challenge in operating an incinerator power plant is to control the burn temperature so that it is high enough to prevent the creation of dioxins from plastics and low enough so that not too many nitrogen oxides are produced. Many incinerators had failed this challenge initially. The health impact of diox-ins includes damage to the central nervous system, the immune system, repro-ductive health system and can impair thyroid function. Because of many of these concerns, local opposition groups proved effective at preventing the construction of new incinerators. By the late 1990s, incinerators had become controversial and politicized, and few new incinerators have been approved since 2000. The "solution" to garbage in US cities is unlikely to be incinerators, but for other cities in countries such as Japan, incinerators are the key method for dealing with solid waste (see Figure 10.5).

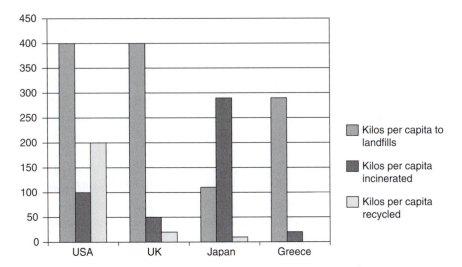

Figure 10.5 *Municipal waste production and disposal, late 1990s*
Source: American Association for the Advancement of Science (2000). **Atlas of Population and Environment,** *Los Angeles and Berkeley: University of California Press, p. 25.*

Solid waste management in Dhaka, Bangladesh

Dhaka, the capital of Bangladesh, is one of the fastest growing megacities in the world. It is a city with a population of over 10 million. It is estimated that some 45–55 percent of the population are poor and live in slums and squatter settlements with little or no access to municipal sanitation services. The city generates about 3,500 tons of solid waste each day, of which 80 percent is organic.

However, only 45 percent is collected and disposed of by the municipality. Informal sector waste pickers, popularly known as *tokais*, collect many of the recyclable materials, such as alumunim and glass. However, they tend to leave the organic material alone and so there remains considerable organic material in the solid waste stream. Often the waste not collected by the municipality or removed by the *tokais* is left to rot in the heat and humidity of the city's open spaces. The resulting stench, rodents and clogged drains pose a serious health risk to Dhaka's residents. In addition, almost half of the waste generated in Dhaka is disposed of in an environmentally unsound way. As the city expands horizontally, it is becoming even more difficult to locate large waste disposal sites within easy access of the city. As a result, the city pays higher transportation costs to haul solid waste longer distances.

Continued

In 1998 a Bangladesh non-governmental organization, called "Waste Concern", initiated a program that promotes the use of organic solid waste for compost to help improve the fast depleting topsoil fertility in rural areas. The city benefits from organic waste recycling through composting because this reduces disposal costs, prolongs the life-span of disposal sites and reduces environmental impacts of landfills. Involving the population in the use of compost programs promotes awareness of garbage while composting activities help create employment and generate income. This innovative partnership includes the government, which makes small parcels of vacant land available free of charge for composting plants, the NGO Waste Concern, which collects, separates and turns solid waste into organic compost and the community, whose members pay a nominal fee for the collection of solid waste, and a private business that markets the organic compost. Currently this program serves 30,000 people in Dhaka and another 100,000 people in 14 other cities in Bangladesh.

One of the interesting aspects of this program is that the composting plants are not highly mechanized, but are small-scale, decentralized, community-based com-posting areas that make use of manual labor. The composting facilities have gen-erated employment for about 16,000 urban poor, particularly women. Consider the story of Makusda, a 32-year-old woman who works in one of the community-based composting plants. Previously she worked in a garment factory, but left because she was often forced to work long hours with no holidays and the salary was very low. There were also lull periods where there was no work in the garment factory and she often went for two or three months without a salary, causing hardship for her family. In the composting factory, she has a 6-day week fixed hours of work with rest time and access to toilet and bathing facilities. She also earns more money than she did at the garment factory. More recently Waste Concern introduced solid waste composting in two of the city's slums by supplying specially designed barrels for composting on site. This not only reduces the waste disposal problem at the source, but also generates income for slum dwellers.

Sources: Adapted from Enayetullah, I. and Sinha, A.H.M.M. (2001) "Public–Private Community Partnerships in Urban Services (Solid Waste Management & Water Supply) For the Poor: the experience of Dhaka City". Bangladesh country report prepared for the Asian Development Bank, Manila; and also

Continued

United Nations Economic and Social Commission for Asia and the Pacific (2002). *Solid Waste Management in Bangladesh.* Retrieved on line at: http://www. unescap.org/rural/bestprac/waste.htm See also Waste Concern. www. wasteconcern.org

The developing world

There are some significant differences in dealing with and providing resources for refuse in the developed and developing world (see Table 10.5 and Table 10.6). In developing countries, individuals living in urban areas use nearly twice as many resources per capita as those living in a rural setting.[210] Because they consume more, they also generate more solid waste. For many countries in the developing world, a lack of resources presents significant obstacles to dealing with solid waste. Inadequate resources limit the ability of cities to collect and dispose of solid waste. Most countries have not been able to upgrade open dumps to sanitary landfills. Few have taken steps to construct, operate or maintain sanitary landfills. Most countries in Africa, for example, continue to practice open dumping. In Brazil, only 10 percent of solid waste is dumped in sanitary landfills; 76 percent is dumped in illegal landfills and another 13 percent in open dumps. Illegal open dumps create risks for the water supply and the health of the urban population. Few monitor the environmental effects of waste disposal. Because many developing countries do not have sufficient resources, most of what they spend is on collection, rather than on maintenance or the upgrading of dumps to sanitary landfills.

How will developing countries deal with increasing waste and the need for more landfills or disposal systems, should rising affluence equate with increased waste generation? Today, cities in Asia generate about 760,000 tons of municipal solid waste per day; by 2025 this figure could increase to 1.8 million tons and they will likely need to double their current $25 billion per year on solid waste management.[211] As a whole, urban populations in the developing world will likely triple their current rate of MSW generation over the next 25 years, while cities in Nepal, Bangladesh, Vietnam, Laos and India might see four to six times the current amount. Such dramatic increases will place enormous stress on already limited financial resources.[212]

There are, however, differences in the waste stream in developed countries to that of developing ones – despite the difficulties in comparisons because what is

Table 10.5 *Comparisons of solid waste management for selected cities*

Activity	Lowest-income cities Lima, Kinshasa, Lahore, Jakarta	Middle-income cities Seoul, Rio de Janeiro, Bangkok	High-income cities London, Frankfurt, New York City, Tokyo
Source Reduction	No organized programs but low per capita waste generation rates	No organized programs	Education programs, some voluntary source reduction programs
Collection	Sporadic and inefficient. Service limited to high visibility and/ or wealthy areas	Improved services and increased collection from residential areas; small vehicle fleet	Collection rate greater than 90%. Highly mechanized vehicles common
Recycling	Recycling through informal sector, waste pickers; localized markets for recyclables	Informal sector still involved; some high technology sorting	Formalized recycling collection services and high technology sorting. Increased attention to developing long-term markets
Composting	Rarely undertaken	Some small-scale composting, but not on large scale	Becoming more popular at large-scale facilities
Incineration	Not common; high moisture content in waste	Some limited use of incinerators	Prevalent in areas with high land costs. More prevalent in European cities than in US
Landfills	Mostly open dumps	Mix of open dumps and sanitary landfills	Engineered sanitary landfills with liners, methane and leachate collection systems
Costs	Collection costs are 80-90% of solid waste budget	Collection costs represent 50-80% of budget	Collection costs represent less than 10% of budget; other $ allocated to landfills, incinerators or recycling programs

Source: Urban Development Unit, World Bank, May 1999. "What a waste: solid waste management in Asia", Washington, DC: World Bank, p.19.

Table 10.6 *Solid waste management cost for selected cities*

City	US dollars per person per year
New York	110
Toronto	70
Caracas	7
Bogotá	8
Buenos Aires	10
São Paulo	15
London	45
Strasbourg	65
Budapest	15
Bucharest	4
Dhaka	Less than 1
Kuala Lumpur	13
Manila	3
Macao	40
Fukuoka	190
Accra	Less than 1
Ulan Bator	Less than 1

Source: Solid waste management cost for selected cities. (2006), in *UNEP/GRID-Arendal Maps and Graphics Library*. Retrieved 16:11, March 13, 2007 from http://maps.grida.no/go/graphic/solid_waste_management_cost_for_selected_cities1

counted as MSW differs from country to country. Typically in rich countries more than one third of waste is paper, plastic accounts for 9 percent and organic materials such as food wastes and yard trimmings account for 28 percent. In the lowest income countries, some 40–85 percent of the waste stream is organic compostable matter, and paper accounts for around 5 percent. Cities in India and China diverge from this trend because they use coal as a household fuel source, thereby generating large quantities of dense ash. The percentage of consumer packaging (plastic, paper, glass and metal) tends to be lower; packaging wastes correlate with the population's degree of wealth and urbanization. However, as countries become richer and more urbanized, they may see a significant increase in paper and packaging – more newspapers and magazine, fast-service restaurants and single-serving beverages.

Informal recycling

The city of São Paulo is the third largest generator of domestic solid waste after New York and Tokyo. Almost 14,000 tons of refuse and 5,000 tons of

Figure 10.6 *Informal trash heap. A sacred cow sits atop trash along the roadside to Agra, India. As in many developing cities, lack of sanitary landfills and collection programs can mean informal dumping of household wastes.*
Source: Photo by Michele A. Judd

industrial solid waste is collected each day at a cost of $150 million and sent to one of three nearby landfills. However, Brazil has a higher rate of recycling compared to the US and Japan. While efforts to privatize recycling have largely failed, the brute fact of high unemployment has created a new economic activity where poor families collect aluminum cans and paperboard. Some 110,000 Brazilians make their living collecting cans on the street and earn an average of 200 dollars per month. These are sold to the industries generally through middlemen and are then recycled. Brazil recycles more than 64 percent of its aluminum cans, 35 percent of its glass, 37 percent of its paper and 12 percent of its plastic.[213] Alcana, a major aluminum manufacturer operates its largest recycling facility in South America in the city of São Paulo. Hence recycling programs in cities such as São Paulo or Rio de Janeiro take place through the informal economy; and ironically, the success of formal recycling programs could eliminate this form of income for the urban poor.

Informal recovery has become a common practice in many developing cities. In Mexico City some 15,000 "scavengers" live in dumps and entire families work to recover cardboard, glass, metals and plastics. Waste picking means higher rates of recycling and provides opportunities for the urban poor to generate some income. However, waste pickers are exposed to severe hazards and can pose a safety threat to themselves and to landfill employees, by interfering with operations at the tipping face, and accidentally starting fires.

Conclusion

In this chapter we have explored how cities are coping with increased levels of municipal solid waste. This form of pollution, perhaps more than any other, is directly linked to levels of affluence and to cultural values that avoid challenging the production of such vast volumes of consumer goods in the first place. For cities in the developing world, many of which cannot cope with either the collection or disposal of garbage, the future may prove even more challenging should these societies achieve the levels of consumption found in the developed world.

Guide to Further Reading

Cozic, C. (1997) *Garbage and Waste*, San Diego: Greenhaven Publishers.

Engler, M. (2004) *Designing America's Waste Landscape*, Baltimore and London: The Johns Hopkins University Press.

Gandy, M. (1994) *Recycling and the Politics of Urban Waste*, New York: St Martin's Press.

Hawkins, G. (2005) *The Ethics of Waste: How We Relate to Rubbish*, Lanham, MD: Rowman and Littlefield.

Johannessen, L. and Boyer, G. (1999) *Observations of Solid Waste Landfills in Developing Countries: Africa, Asia and Latin America,* Washington, DC: The World Bank.

Ludiwig, C., Hellweg, S. and Stucki, S. (2003) *Municipal Solid Waste Management: Strategies for Sustainable Solutions,* Berlin: Springer.

Newsday. (1989) *Rush to Burn: Solving America's Garbage Crisis*? Washington, DC and Covelo, CA: Island Press.

Rathje, W. and Murphy, C. (1992) *Rubbish! The Archeology of Garbage*, New York: HarperCollins.

Royte, E. (2005) *Garbage Land: On the Secret Trail of Trash*, New York: Little Brown.

Strasser, S. (2000) *Waste and Want: A Social History of Trash,* New York: Owl Books.

Williams, P.T. (2005, 2nd edn) *Waste Treatment and Disposal*, Chichester: John Wiley.

Part III
(Re)aligning Urban-Nature Relations

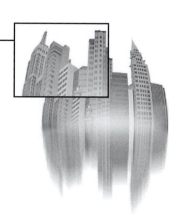

11 Race, class and environmental justice

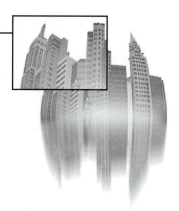

The city is both an environmental and a social construct. It is predicated upon ecological processes, indeed it is a complex ecological system in its own right, yet it is also a social artifact that embodies and reflects power relations and social differences. The city is at the center of a social–environment dialectic that connects the environmental and the political. In this chapter we will explore how issues of class, race and gender interconnect with environmental issues.

Urban environments and race

It is a consistent finding that toxic facilities are concentrated in minority-dominated areas of the city, and major infrastructure projects with negative environmental impacts such as urban motorways are more commonly found in minority neighborhoods. Study after study reveal a correlation between negative environmental impacts and the presence of racial/ethnic minorities.[214] An early study in the US revealed that race was the most significant variable associated with the location of hazardous waste sites and that the greatest number of commercial hazardous facilities were located in areas with the highest composition of racial and ethnic minorities. The study also showed that three out of every five Black and Hispanic Americans lived in communities with one or more toxic waste sites. Although socioeconomic status was also an important variable in the location of these sites, race was the most significant.[215]

Take the case of Chester, PA, a typical industrial town outside the city of Philadelphia (see Figure 11.1). It grew as a manufacturing center with steel mills, shipyards, aircraft engines plants and a Ford Motor Company plant. By the 1970s, however, deindustrialization had begun to erode much of the manufacturing base. As firms closed, workers left and the town's population became increasingly poor, older and black. By the time of the 2000 Census the population

Figure 11.1 *Chester, Pennsylvania. This former industrial city, with a predomi-*
nant African American population, has been the site for the location of several
waste incinerators since the 1980s.
Source: Photo by John Rennie Short

of 36,851 was 75 percent black compared to the state average of 10 percent, and
the median household income was $13,052 compared to the state average of
$20,880. A quarter of the town's population lived below the poverty line.

Environmental racism and injustice?

Just because a hazardous site is situated in a minority community does not necessarily mean that environmental racism is at work. Consider the case of trash
transfer in Washington, DC. The city needed sites to consolidate its garbage collection and disposal system, the narrow city streets meant that only relatively
small trucks could collect garbage in the city, but larger trucks were needed to
move the trash to remote incinerators and landfills. A plan emerged from an
independent committee in the 1990s to build a new transfer station in Ward 8
in the city. The site was ideal; it was flat, the city owned the property, it had good
accessibility for the truck traffic and was large enough to satisfy a city ordinance
requiring a 500 foot buffer zone between transfer stations and residences. The new

Continued

proposed site also allowed the closure of older, noisier and more polluting transfer stations in other minority neighborhoods. But Ward 8 houses a predominantly low-income minority population. In the lively debate that followed the terms "environmental racism" and "environmental injustice" were often used. A closer reading of the facts, however, show that there were few other options. Local activists could easily use the widely known terms in the polarized racial politics of Washington, DC; it was an obvious and easy rhetorical point to make. The terms environmental racism and environmental justice can be powerful words of rhetoric even when they have little explanatory purchase.

Source: Talking Trash http://www. Washep.com/archives/cover/2001/cover0126.html (accessed June 10, 2006)

Eager to lure tax-generating facilities, the city government in the 1980s sought to redevelop old factory sites, attract business and generate jobs and tax revenue. Chester provided an ideal opportunity for certain industries as it was poor, desperate, and had land, while local communities had limited political power. Other cities and areas might have provided stronger resistance to polluting enterprises. A large real estate developer bought up the land rights in the old industrial area and leased space to other businesses. In 1987 the Pennsylvania State Department of Environmental Protection granted permits for three waste facilities in the city of Chester and two outside the city. By the mid 1990s the city housed the nation's largest concentration of waste facilities including a trash transfer business, an incinerator, a medical waste sterilizing facility, a contaminated soil burning facility, a rock crushing plant and wastewater plant that handled effluent from factories, and a refinery. A local citizen's group, The Chester Residents Concerned for Quality Living, claimed that the toxic emissions led to low birth rate babies and local cancer clusters. They also claimed environmental racism, taking their case against the state Department of Environmental Protection (DEP) to the courts. They argued that between 1987 and 1996 the DEP approved permits for 2.1 million tons of landfill in black areas of the city, but approved only 1,400 tons in white areas. A federal judge threw out the lawsuit in 1996 stating that there may have been a discriminatory *effect* but the residents could not prove a discriminatory *intent*. A court of appeal reversed the judge's decision but the Supreme Court dismissed the case in 1998. However, by even agreeing to hear the case at all the Supreme Court signaled the viability of environmental racism as a legal argument.

In Chester there was a cluster of noxious, polluting facilities in predominantly poor, black, residential areas. This is an extreme case but one that highlights the character of environmental racism. It is less a formal legal issue since the permits for the facilities had been issued correctly and formal procedures had been followed, and represents a more moral issue. Vulnerable and poor residents were dumped on both metaphorically and literally.

It is often difficult to untangle the intent and effects of environmental racism. The term "environmental racism" implies both, but in the case of Chester, while the effect was obvious the intent was more difficult to prove. In counter-arguments it was argued that the city was so poor that it needed to attract tax-paying operations to fund city services. The rationale for this cost-benefit may be unbalanced; the health of residents far outweighs the contribution to the city's coffers. However, the simple causal processes implied by the cavalier use of the term racism need some caution.

Environmental negative impacts occur in racially homogeneous nations. In Japan, for example, there are 2,000 highly polluting municipal incinerators that spew out a deadly cocktail of heavy metals, nitrous oxides, carbon dioxide, toxins and furans that lead to cancer, birth defects and illness. The results are spread throughout the cities of Japan, spewing pollution on rich and poor Japanese.

In a careful study of environmental inequity, Christopher Boone and Ali Modarres examined the case of Commerce. This predominantly Latino city east of Los Angeles has a high concentration of polluting manufacturing plants. They show that the businesses located in the city because it provided vacant land and accessibility. The bulk of industry was located in the city before the demographic changes that made it a Latino city. In other words, there were toxic neighborhoods because of factors of accessibility and land availability more than because they were Mexican neighborhoods. The Boone and Modarres study suggests that immigrant, migrant and minority communities may develop around toxic areas because of the operation of the housing market, where the poorest people get the least choice and end up in the worst neighborhoods, rather than the case that toxic areas are knowingly located in minority neighborhoods.[216]

This is not to dispute the connector between environmental quality and minority residential areas. In the US, among counties that have three or more pollutants 12 percent are majority white, 20 percent are majority African American and 31 percent are majority Hispanic. Race and ethnicity intertwine with issues of power and access to power to produce an uneven experience of environmental quality at home and in the workplace.

While the connections between intent and effect are sometime difficult to disentangle, race and ethnicity can play a role in mobilization. Robert Bullard provides case studies of community disputes in cities in the US ranging from conflict over solid waste landfill in Houston, a lead smelter in Dallas and a solid waste incinerator in Los Angeles. Race can become a site of mobilization, a shared experience on which to build resistance and to fight against environmental injustice.[217]

Urban environments and socioeconomic status

Socioeconomic status plays a major role in the environmental quality of urban living. Poorer communities have less pleasant urban environments and often bear the brunt of negative externalities. It is through their neighborhoods that motorways are constructed; it is in their neighborhoods that heavy vehicular traffic caused, and in some places still causes, high lead levels in the local soil and water. There is a direct correlation between socioeconomic status and the quality of the urban environment.[218]

The causal web is sometimes complex but often simple. Poor people get dumped on because they are poor, and they are poor because they lack the wealth to generate political power and bargaining strength. The city is a space where the best areas go to those with the most money and those with the least get what is left. On top of this historical relationship there is the current trend in the siting of noxious facilities that tends to skew their location to those with least power to resist. The poor are the least powerful and experience the worst urban environments. In more racially homogeneous cities the deciding factor is socioeconomic status.

In some cases, the correlation between race and class on the one hand and environmental quality on the other was, and in some cases still is, reinforced by political movements and economic forces that tended to discount environmental quality in favor of economic growth and employment opportunities. Many labor movements were slow to realize that environmental issues were social justice issues, not just the superstructural concerns of the affluent. And the brute economic forces of the industrial city often forced a false divide between environmental qualities versus jobs. Crenson, for example, tells the story of the lack of an environmental movement in many industrial cities in the US because of the supposed linkage between pollution and employment. Smoking chimneys signified good well-paying jobs. We have also highlighted this issue in our discussion of Onondaga Lake in Syracuse. For years the business elite promoted the issue in terms of "Does Syracuse want its people employed, or do they want the lake cleaned up?" It was only in the 1970s that a cleaned-up lake was reimagined as a vital part of a new postindustrial city. But as long as it was an industrial city, lake cleanup was a distant second to economic growth. And even today in many cities, especially in the developing world, dictates of business expansion and economic growth often outweigh the environmental concerns of, and the quality of urban life for, the majority of the people.[219]

Racialized topographies

Richmond, Virginia, whose residents in 2000 numbered 818,836, of whom 32.4 percent were African American, had a socio-spatial pattern of lower-lying, inner-city, black areas and surrounding white-dominated hills. The correlations between percent black and altitude were -0.41 in 1990 and -0.47 in 2000. Richmond has a long history of segregating races by elevation. It was one of the first southern cities to embrace racial zoning in 1910. Richmond's local elites used the state to control the expansion of black communities. African Americans were crowded into the dilapidated houses in the Jackson Ward area. East End and Church Hill absorbed a large influx of blacks after World War II. Residential discrimination extended into public policy decisions in the placement of public housing. Sixty-four percent of public housing units built prior to 1970 were placed in Church Hill and Jackson Ward. These decisions reinforced the racial topography of Richmond.

Source: Ueland, J. and Warf, B. (2006) "Racialized topographies: altitude and race in southern cities", *Geographical Review* 96(1): 50–78.

We can now see the poverty of this false dichotomy of jobs versus environment. Low-grade environments have their most negative health impacts on working people. It is not a case of jobs versus the quality of the urban environment but jobs *and* the quality of the urban environment. As we move further into a greener economy there are direct connections between improving the urban environment and employment opportunities. Recycling, green technologies and greening the city are all ways to create jobs in hard-pressed cities.

Environmental justice

In 1994 President Clinton signed Executive Order 12898 that initiated an environmental justice program within the Environmental Protection Agency (EPA). The aim was to raise awareness of environmental justice issues, identify and assess inequitable environmental impacts and provide assistance to local areas and community groups.

Environmental justice issues arise from the obvious fact that there is a correlation between the siting of hazardous facilities and low-income communities and/or minority communities. However, as the case of Chester shows, legal redress is difficult if procedures were correctly followed. Environmental justice, like social justice, is not possible in the absence of more interventionist methods. The normal workings of a racist, classist society will produce racist, classist outcomes in the normal course of events, even without the aid of illegality or the help of corruption. In order to produce more equitable outcomes we need more positive interventions. Environmental impact statements, for example, need a more explicit assessment of equity and justice issues. Low-income communities facing environmental challenges should also receive greater resources from government, not simply equal treatment under the law. Unless more positive outcomes are engineered, the system will tend to produce inequitable results.

Urban environments and social difference

The experience of living in the urban environment varies across the dimensions of social difference. Socioeconomic status, gender, age and level of physical ability/disability are all sources of social difference that are embodied and reflected in urban environments.

Take the case of gender. Women have played an important role in bringing environmental issues to a broader public. Lois Gibbs, for example, was a typical suburban housewife with two children living in a suburb in western New York. When her son Michael developed epilepsy, her daughter Melissa contracted a rare blood disease and her neighbors' children also got sick, she began to look into possible causes. She and others soon discovered that the area where she lived, Love Canal, was built on a toxic waste site, with more than a dozen known carcinogens including the deadly chemical dioxin. The local soil and water was poisoning the community. Lois Gibbs became one of the community organizers mobilizing public opinion and promoting state and federal involvement. The area was evacuated in 1980, the area's name joining places like Bhopal as a byword for environmental pollution. The publicity given to Love Canal strengthened the case for the Superfund legislation, in which chemical cleanup of major toxic sites was federally mandated, which was passed in 1981.[220]

The experience of urban toxicity is often experienced by women in households and neighborhoods where women play a strong role in household care and management. They tend to children's illnesses, connect with local neighbors and may be closer to local issues. It is often women who are on the frontline of local environmental issues.

The case of Love Canal in which children became very sick is also an example of the effect of age on the experience of urban environmental conditions. Embryos in the womb and very young children are especially vulnerable to adverse environmental conditions. They are the human equivalent of the canary in the mine, providing early and tragic warnings of the state of our environment. Older people are also especially vulnerable to air quality and other environmental conditions. Yet in many cities the very young and very old have the least voice but are often the most impacted. Most cities are designed by and for affluent males; the further from this profile the more marginal to most discourses of power and influence. The worst urban environments are experienced by the more marginal; the poor, young and old.[221]

Healthy city initiatives

One obvious healthy city initiative is to encourage more vegetation growth. A careful tree-planting program, for example, can lower summertime temperatures, minimize the urban heat island and reduce air pollution. Some cities actively encourage tree planting. In Sacramento, CA, since 1990, over 375,000 shade trees have been given away to city residents with plans for 4 million more trees to be planted throughout the city. In other US cities, in contrast, tree-planting programs have been cut back. A study of 24 cities in the US showed there has been a 25 percent decline in tree canopy in the past 30 years. In cities such as Milwaukee the tree-planting program was reorientated away from communities to greening the downtown to make it more attractive to investors. The Sacramento experience, however, shows that tree planting not only makes for a healthier city, it also saves money. The city estimates that for every dollar spent on trees it recoups $2.80 in energy savings, pollution reduction, stormwater management and increased property values.

There has been resistance to tree planting from some utility companies who argue that it costs more to secure their power lines though vegetation. In states such as Iowa, in contrast, which mandates tree planting by utility companies, the experience has been more positive. In a time of rapid energy costs, utility companies are often grateful for the public relations bonus they receive from partnering in tree planting.

Source: Harden, B. (2006) "In California city, shade crusade takes root to cut energy costs", The Washington Post, September 4: A and A10.

Healthy cities

In 1984, the World Health Organization sponsored a conference in Toronto entitled *Healthy Cities*. Two years later, the Ottawa Charter for Health Promotion outlined the basic requirement of improving overall health by improving the physical and social environments of cities. There are now almost 2,000 healthy city initiatives around the world. New initiatives have been promoted along with existing public health programs under the term healthy cities. In Europe, the emphasis has been on integrating health and urban planning. In Latin America, healthy cities are more associated with the provision of basic infrastructural services of water and sewerage. There has been a more explicit acceptance that healthy cities must connect to issues of environmental justice. In 2001, member counties of the EU ratified the Aarhus Convention that every person has the right to live in an environment adequate to his or her health and well-being, and to achieve this end citizens must be involved in decision making and have access to information. The Convention promoted early and effective participation and various evaluation criteria. Hartley and Wood examined public participation in the environmental impact assessments in four UK waste disposal case studies in the wake of the Aarthus Convention. They report that the Aarhus Convention led to a strengthening of participation procedures but that the level of improvement secured would depend upon how its ideals were interpreted and incorporated into specific legislation. In other words, it is the legislative and administrative details that will structure the implementation.[222]

Although the pronouncements of the Aarhus Convention are easy to make, rhetoric being easier than real action, they do hint at the future connections between issues of social justice, participatory democracy and urban environmental quality. They also indicate that environmental rights may join the list of citizen rights and government obligations that mobilize communities and motivate governments.

Guide to Further Reading

Adamson, J., Evans, M. M. and Stein, R. (eds) (2002) *The Environmental Justice Reader,* Tucson: University of Arizona Press.

Bullard, R. D. (ed.) (2005) *The Quest for Environmental Justice: Human Rights and the Politics of Pollution*, San Francisco: Sierra Club.

Bullard, R. D. (ed.) (2007) *Growing Smarter: Achieving Livable Communities, Environmental Justice, and Regional Equity,* Cambridge, MA: MIT Press.

Pellow, D. N. and Brulle, R. J. (2005) *Power, Justice and the Environment: A Critical Appraisal of the Environmental Justice Movement,* Cambridge, MA: MIT Press.

Short, J. R. (1989) *The Humane City,* Oxford: Basil Blackwell.

Stein, R. (ed.) (2004) *New Perspectives on Environmental Justice: Gender Sexuality and Activism*, New Brunswick: Rutgers University Press.

Takanos, T. (ed.) (2003) *Healthy Cities and Urban Policy Research*, London: Spon.

Tsourous, A. (2006) *Healthy Cities in Europe,* London Spon.

Washington, S.H. (2005) *Packing Them In: An Archaeology of Environmental Racism in Chicago,* Washington, DC, and Covelo, CA: Rowman and Littlefield.

12 Sustainable urban development

The cities of the twenty-first century are where human destiny will be played out, and where the future of the biosphere will be determined. There will be no sustainable world without sustainable cities.

Herbert Girardet[223]

In the previous chapters we have looked at environmental issues in an urban setting focusing on perceived problems and possible solutions. In this, the final chapter, we want to consider how cities can be and are being managed in order to minimize negative environmental impacts and to promote sustainability. We focus on reactions to growth, the rediscovery of place and urban sustainability. These discourses ask us to fundamentally rethink and redesign existing cities.

Dealing with growth

Urban growth has been a pervasive feature of the past hundred years. In cities large and small and in countries rich and poor, growth has been a dominant feature. Cities that once occupied only a few square miles, now cover hundreds. Urban residents who once walked most places now depend on the automobile. Suburban sprawl has increased as metropolitan areas extend their reach further out into the countryside. The metropolitan sphere of influence deepens its shadow across the landscape as farmland turns into tract housing, woods into subdivisions and the prairie into gated communities. In this section we will consider various responses to dealing with urban growth. We can characterize five main responses as *resistance, smart growth, new urbanism, slow growth and historic preservation*.

Resistance

Resistance occurs at all scales. It occurs especially in growth areas with more affluent households. In the high growth areas, rapid increase in traffic, overcrowded schools and local property taxes can all stimulate local resistance. In Howard County in Maryland, for example, where population grew 32 percent from 1990 to 2000, and a further 7.6 percent to a 2004 population of 266,738, some residents began to resist the latest round of development proposals. The powerful development house-building lobby marshaled against such resistance. When residents in Howard County signed a petition to create a referendum to challenge rezoning proposals, a group of landowners sued the council to cancel the referendum.

Resistance is particularly strong when the local residents are wealthy and organized. Mark Singer describes a struggle in a wealthy Connecticut town as one between the haves and the haves. The town of Norfolk, CT, is a place of old money and obvious displays of affluence are frowned upon. Almost 80 percent of the town's 30,000 acres is designated as forest, agriculture or park. An area of 780 acres, known as Yale Farms came on the market in 1998. A plan to develop a luxury golf course and one hundred homes, each on four acres, generated intense resistance from groups that called themselves the Canaan Conservation Coalition and the Coalition for Sound Growth. The fight, according to Singer, has given people a chance to affirm their shared values. Behind the debates about loss of green space and loss of community was fear of change and a distaste for the incoming nouveaux riches who had enough money to disrupt the traditional moral code with their ostentatious wealth.[224]

Resistance takes many forms from "not here", "not this here", to "not this here now". The success of the resistance depends upon the wealth, organizational skill and effective links to political power of the pressure groups. But the battle is uphill and in many cases unsuccessful. The resistance movement has to compete with the powerful development, real estate and property investment interests. Successful resistance is not a global phenomenon. In some parts of the world citizens lack access to power and political leverage declines with income and status. In China, for example, which has experienced arguably the highest rates of recent urban growth, rural peasants and urban dwellers have little recourse in the face of state-prompted urban development.

Smart growth

One planning response to sprawl is the *smart growth* initiative that stresses mixed land uses and compact building designs that create high densities with lower environmental impact. Smart growth has emerged as a strategy to deal with the constant pull of development toward greenfield sites on the city's edge; it focuses on existing developments in order to utilize their infrastructures and to preserve open space and farmland. It is a framework for municipalities facing heavy development pressure and looking for principles and policies to halt the abandonment of urban infrastructure and the costly rebuilding in greenfield sites. In 1996, the Smart Growth Network enunciated some principles: mix land uses, design more compact buildings, construct walkable communities, create a sense of place, preserve open space, direct development toward existing communities, provide a variety of transport choices, make fair, predictable and cost-effective decisions, and encourage community involvement in development decisions. Smart Growth argues the remedy for sprawl is to create housing development dense enough to encourage public transportation and to reduce the need for the automobile.

In 1997, the state of Maryland in the US established a set of smart growth policies with three main goals: to target state resources to areas where infrastructure was already in place; to preserve farmland and natural resources; and to resist public investment in building infrastructure that promoted sprawl. A set of priority funding areas was identified for state funding of transportation, water and sewage. In effect, the policy guided higher-density development in areas already served by public infrastructure. The Maryland smart growth initiative under the leadership of Governor Parris Glendenning was a model for other states. The statewide scheme was abandoned in 2002, with the election of Republican Governor Robert Ehrlich, but at the county level a variety of smart growth strategies continues. In Montgomery County MD, for example, planning proposals include the encouragement of high-density mixed land use, the use of infill developments and plans for more development at bus, metro and rail stations and the transformation of automobile-dominated roads into boulevards.

Smart Growth is a possible answer for municipalities facing heavy development pressure and looking for principles and policies to halt the abandonment of urban infrastructure and the costly rebuilding in greenfield sites. But it is still too early to say whether smart growth will become an effective policy to halt the seemingly relentless expansion of the suburban fringe into open spaces.

Smart growth in combating sprawl

The chief development alternative to emerge in response to sprawl is "smart growth". With its focus on urban revitalization and expanded transit options, smart growth seeks to make existing communities places where people want to live.

Continued

The term was popularized by Parris N. Glenderning, governor of Maryland from 1994 to 2002, who in 1997 launched the Smart Growth and Neighborhood Conservation Program to limit sprawl in his state. Today dozens of environmental groups, civic organizations and government agencies promote smart growth principles as part of their sprawl reduction programs. These principles include, among other concepts, the promotion of mixed land uses and the creation of attractive neighborhoods with a strong sense of "place" or local identity and character, where residents can walk freely to the places they need to go.

Portland, Oregon, is a well-known model of sprawl containment. The city established an "urban growth boundary" in 1980 that protects nearby farmland surrounding the city and tightly limits development in outlying areas. Portland's approach has not been without controversy. For several years, the urban growth boundary was accompanied by skyrocketing housing costs and discontent among those who resented restrictions on development. But the high costs of housing – which are in fact attributable to a host of factors, including a high rate of migration to Portland from other states, particularly California – have since declined to the point that they are roughly equivalent to those of other West Coast cities.

Because of the urban growth boundary, Portland has assimilated a sharply rising population without encroaching on its valuable land resources. Portland's urban designs provide affordable and accessible public transport located close to schools, businesses and residential communities. In addition, walking and bike paths connect the entire community, which is infused with a multitude of parks and green spaces.

Source: Schmidt, Charles W. (2004.) "Sprawl: the new manifest destiny?", *Environmental Health Perspectives*, 112 (11): p. A620 – A627.

New urbanism

The history of urban planning is full of attempts to reorganize the city along principles of rational efficiency, good design and encouragement of community. The latest in a long line of urban design movements is New Urbanism. It is a response to suburban sprawl that emphasizes revitalizing old urban centers; creating mixed-use centers where residences are located close to commercial and office sectors; planning for walkable, high-density, low-rise residential areas that are socially diverse communities; minimizing the speed of autos through urban areas and making cities more attractive to walking and casual social interaction.[225]

There is not anything particularly "New" about these ideas; the century-old Garden City Movemend promoted most of these ideals.

One example of the New Urbanism is Seaside in Florida, designed by Andres Duany and Elizabeth Plater-Zyberk, where neo-traditional houses conform to a strict code, the houses are at a high density, car traffic is kept to the edges and the walkways and porches are constructed in order to aid community interaction. Duany in particular has become a high profile advocate of New Urbanism. Seaside was the community shown in the movie *The Truman Show*.

Perhaps the most cited example of New Urbanism is Celebration, a community initially planned and funded by the Disney Corporation. Walt Disney originally had a plan for an Experimental Prototype Community of Tomorrow (EPCOT). However, EPCOT was integrated instead into Florida's Walt Disney World theme park and was never realized as a community. Celebration opened in 1996 just outside Orlando, Florida. With its Charles Moore civic center and Ceser Pelli cinema, the place has attracted big name architects and lots of attention.[226] There are now almost as many books and articles about Celebration as residents (just over two thousand lived there in 2000 with a planned population of 20,000). Celebration has all the design elements of the New Urbanism: low-rise, high density residential areas where garages are at the back of the residences, walkways and porches allow pedestrian movement and these is a mixed-use downtown. It is also emblematic of the exclusivity that has characterized New Urbanism. The lowest rents are $800 per month, and most of the population is upper-income. While the New Urbanism proclaims social heterogeneity, in practice it tends to be restricted to the middle- and upper-income groups.

In Europe, the equivalent of New Urbanism is the "urban village". In Copenhagen, Freiburg, Vienna, Zurich, Heidelberg and Barcelona, urban redesign has focused around pedestrian streets, sidewalks, public squares and parks. The dense pedestrian center is then connected to the rest of the city with bicycle lanes and well-integrated public transport.[227] The result is more walkable cities that offer diverse attractions that aim to foster a sense of urban community.

Underlying the New Urbanism and the urban village model is a nostalgic sense of community and neighborliness, a longing for a lost community. The older high-density cities of the half-remembered, half-created past, are often portrayed as places of tight community, while more recent suburban growth is seen as a cause of the decline of the community. The New Urbanism is a catchall phrase that in principle captures the discontent with contemporary developments, especially the nature of low-density sprawl and the alienation felt by many residents.

In practice, it means developments that are high-density, pedestrian-friendly and socially exclusive.

The New Urbanism is not all that new and is not all that urban. On closer inspection it looks like the latest version of up-market suburban communities. New Urbanism, as it has been practiced so far, does little to discourage suburban sprawl since it still produces densities too low to support public transportation and truly mixed communities, creating instead homogeneous enclaves. It is repackaged urban sprawl, a useful marketing strategy, playing an important role in stimulating debate, but with little practical effect on creating community.[228]

Few would argue the need for some kind of alternative to the standard forms of suburban sprawl. It is wasteful of resources, lacks aesthetic appeal and produces a series of edges rather than centers. The New Urbanism is, at the very least, proposing alternatives to a city dominated by the auto, the highway and the parking lot. The New Urbanism as a design guide is a step in the right direction. However, the claims made for the New Urbanism as a source for "recovery of community" are based on a series of assumptions (community is declining, urban form can resuscitate this community) that are asserted rather than demonstrated. The New Urbanism is a source of interesting design ideas, but as a method of recreating community, the jury is still out.

Slow growth

The Slow City movement has member cities in Italy, Germany, Norway and England. It was founded in October 1999 in Italy by three mayors and was designed as a policy framework and network of small cities – they have to have populations of less than 50,000 – seeking to connect the three Es of environment, economy and equity. There are now around 20 cities with a designated list of slow growth environmental policies and urban designs. A study of two German cities in the network points to the protection of city-owned pastures and apple trees and revitalization of houses as community public spaces.[229] Cities in the slow growth network tend to be not only small but also homogeneous with a shared political agenda. In larger more heterogeneous cities with a less interventionist political culture, slow growth may be politically untenable. However, it is suggestive of how groups of cities can network their ideas and share practices of urban sustainability.

Preservation

There has always been some appreciation of cultural and architectural heritage, but only in the last few decades have societies become increasingly aware of the significance of urban historic structures and sites. In the 1950s and 1960s the prevailing attitude was that "*old*" was bad and "*new*" was superior.

In the US the preservation movement evolved from two distinct paths.[230] The *private sector* path focused on important historical figures and landmark structures. It has been called the "George-Washington-Slept-Here" approach. The *public sector* path was involved with establishing national parks and this also included historic buildings. In the 1930s and 1940s, the public sector established some historic districts, such as Charleston, South Carolina, and the Vieux Carré (French Quarter) section of New Orleans and Alexandria, Virginia. In 1949, several organizations evolved into the National Trust for Historic Preservation, inspired by the British version. The purpose was to link the preservation efforts of the private side with the federal government/National Park Service activities. The most important piece of historic preservation legislation was the National Historic Preservation Act of 1966. It established new laws, authorized funds for preservation activities, and encouraged locally regulated historic districts. It went beyond merely protecting landmarks to recognizing a variety of historically and architecturally significant buildings, sites, structures, districts and objects. In addition, preservation no longer focused on saving single landmarks, instead entire areas were delineated as historic districts and became an important tool of urban revitalization during the 1970s and 1980s. Today there are thousands of local preservation associations and thousands of designated historic sites, buildings and other structures in cities. Historic preservation is composed of a variety of strategies: preservation, restoration, reconstruction and rehabilitation.

Preservation refers to the maintenance of a property without significant alteration to its current condition. When preservation is the guiding strategy, the only intervention is normal maintenance or special work needed to protect the structure against further damage. An example of innovation in preservation is Pike Place Market in Seattle. The old city market was threatened with demolition to make way for an urban renewal project. In a city-wide vote, however, residents voted to save the market as an important part of their city's life and culture. In order to prevent the loss of its original character as a working everyday market run by local farmers, fisherman and small entrepreneurs, the city developed an ordinance that not only protected the structure, but also the activities within.[231]

Restoration refers to the process of returning a building to its condition at a specific time period. However, this often means changing the natural evolution of a building, and creating a more contrived picture of its "original condition".

This aspect of preservation is more common in historic homes, farms or churches. This strategy, however, is not without criticism, as it casts doubt on the authenticity of restored structures.

The term reconstruction refers to the building of a historic structure using replicated design and/or materials. This approach is taken when a historic structure no longer exists. The earliest and best example is Williamsburg, Virginia. In 1926, John D. Rockefeller was persuaded to fund the restoration of the entire colonial town of Williamsburg. The primary problem was that much of the original town had been lost over the centuries, and while many of the historic buildings remained, a few central buildings from the town's original layout were missing. Planners decided to reconstruct the Governor's Palace (which had been

Figure 12.1 *Musée d'Orsay in Paris. This former railway station has been converted into an art gallery, an example of adaptive reuse.*
Source: Photo by John Rennie Short

destroyed by a fire in 1781). The efforts to reconstruct Williamsburg were not without controversy – as some buildings were removed to make way for reconstructed ones.[232] Yet it remains one of the most visited historic districts in the US. In addition to both restoration and reconstruction, Colonial Williamsburg presents live recreations of historic events by actors in period costumes. This way of presenting historical places and artifacts, often referred to as a "living history museum", has become increasingly popular.

Lastly, many buildings no longer perform their original function or use but retain their architectural integrity. For these structures, a common strategy is rehabilitation, sometimes referred to as adaptive reuse. The purpose is to modify or update portions of the structure and adapt the building for a new purpose. Numerous examples abound – from abandoned factories that are adapted into microbrewery pubs or museums or residential lofts. Increasingly rehabilitation is a strategy employed by cities seeking to revitalize old areas of the city. Figures 12.1 and 12.2 show the adaptive reuse of former industrial areas. In preserving and protecting urban buildings, structures, and parks and public squares, cities have

Figure 12.2 Docklands, London. This former warehouse in the old industrial area of Docklands, London, has been converted into shops, restaurants and offices. Source: Photo by John Rennie Short

an effective strategy to encourage a return to the center while preventing the obliteration of areas by development.

Historic Preservation is increasingly at the center of urban redevelopment efforts as cities search for a way to celebrate their past while looking to the future. It can be used to preserve neighborhoods as well as ecologically important areas within the city.

Urban sustainable development

Cities are now considered unsustainable in that they cannot continue in the same way – consuming resources and generating waste. During the 1990s a variety of debates about urban growth and urban sustainability included environmental activists, economists, ethicists and planners.[233] What does sustainability mean for cities? There are a variety of terms, adjectives and meanings associated with "sustainability" from ecological, environmental, social, political, to cultural and economic. The concept in part stems from an appreciation that the environment is of limited capacity. It questions the principal Enlightenment project of the mastery of nature.

The concept of sustainable development has emerged from a convergence of intellectual perspectives, as highlighted in Table 12.1. Despite several decades of discussion, no one definition of sustainable development has emerged.[234] The most widely used is "development that meets the needs of the present without compromising the ability of future generations to meet their own needs".[235] Other definitions include "living within the carrying capacity of supporting ecosystems" or "preservation of ecological and social capital" and "that which improves the long-term health of human and ecological systems".[236] International discussions and debates about sustainability began with Agenda 21, the primary outcome of the 1992 Rio Earth Summit. For cities, the Aalborg Charter of 1994 outlined sustainability objectives for European cities (see Box Insert). Subsequent international conferences such as the Istanbul UN City Summit of 1996 and the Local Government Declaration to the 2002 UN Johannesburg Earth Summit expressed similar ideas about directing urban growth in more sustainable ways.

Sustainability has as its guiding agenda the protection of both the local and the global commons – the biosphere and the atmosphere. This means that for cities to embark on sustainable urban development, they need to address not only local environmental impacts, but the urban impacts on the global climate, biodiversity and energy use. Sustainability takes place at a variety of geographical scales,

Table 12.1 Intellectual contributions on Urban Sustainable Development

Environmentalists	Economists	Urban studies/urban planning	Equity advocates	Spiritual writers and ethicists
John Muir (preservation)	Kenneth Boulding	Patrick Geddes, Ebenezer Howard, Lewis Mumford (comprehensive planning)	Murray Bookchin (social ecologists)	Gary Snyder, Thomas Berry, Dalai Lama
Rachel Carson	Herman Daly (steady state economics)	Norman and Susan Fainstein, Robert Beauregard (critique of rational planning)	Edward Goldsmith, Nicholas Hildyard, Frances Moore Lappe, Arturo Escobar, Vandana Shiva and Martin Khor (development critics)	Charlene Spretnak, Petra Kelly, Carolyn Merchant (ecofeminism)
Donella Meadows	Michael Redclift and David Pearce (environmental economics)	John Logan, Harvey Molotch, Brian Stoker, Peter Calthorpe (urban growth coalitions)	Robert Bullard, Carl Anthony (environmental justice)	Baird Callicott, (environmental ethics)
Brundtland Report	Robert Costanza, Richard Norgaard (ecological economics)	David Gordon, Timothy Beatley (Green Urbanism)	Eric Swyngedouw, Matthew Gandy (political ecology)	Theodore Roszak (ecopsychology)
EarthSummit/ Agenda 21	Paul Hawken (restorative economics)	Michael Hough, Rutherford Platt (urban ecology)		Yi-Fu Tuan (topofilia)
President's Council on Sustainable Development (USA)	William Rees (ecological footprint analysis)	James Kunstler, Edward Relph (placelessness)		
Bill Devall/George Sessions (deep ecology)				
Kirkpatrick Sale (bioregionalism)				

Source: Adapted from Wheeler, S. (2004) Planning for Sustainability: creating livable, equitable and ecological communities, London and New York: Routledge p. 28.

including – but not limited to – the city. In this section, we explore how concepts of urban sustainability can be realized within cities at several scales – individuals, neighborhoods, suburbs and cities. We conclude with several examples where sustainable development has been part of an overall urban plan or redevelopment project.

European cities and sustainability

In 1994, The European Sustainable Cities & Towns Campaign was launched in Aalborg, Denmark. The participants at this first European conference discussed and adopted the Charter of European Cities and Towns towards Sustainability (the Aalborg Charter). To date, more than 2,000 European local and regional authorities (metropolitan areas, cities, towns, counties, etc.) from numerous European countries have signed up to the Aalborg Charter. Below are several principles in the Aalborg Charter.

Part I.1 The Role of European Cities and Towns

We, European cities & towns, signatories of this Charter, state that in the course of history, our towns have existed within and outlasted empires, nation states, and regimes and have survived as centres of social life, carriers of our economies, and guardians of culture, heritage and tradition. We have learnt that present levels of resource consumption in the industrialised countries cannot be achieved by all people currently living, much less by future generations, without destroying the natural capital.

We are convinced that sustainable human life on this globe cannot be achieved without sustainable local communities. Local government is close to where environmental problems are perceived and closest to the citizens and shares responsibility with governments at all levels for the well-being of humankind and nature. Therefore, cities and towns are key players in the process of changing lifestyles, production, consumption and spatial patterns.

I.3 Local Strategies Towards Sustainability

We are convinced that the city or town is both the largest unit capable of initially addressing the many urban architectural, social, economic, political, natural resource and environmental imbalances damaging our modern world and the smallest scale at which problems can be meaningfully resolved in an integrated,

Continued

holistic and sustainable fashion. As each city is different, we have to find our individual ways towards sustainability. We shall integrate the principles of sustainability in all our policies and make the respective strengths of our cities and towns the basis of locally appropriate strategies.

1.6 Urban Economy Towards Sustainability

We, cities & towns, understand that the limiting factor for economic development of our cities and towns has become natural capital, such as atmosphere, soil, water and forests. We must therefore invest in this capital. In order of priority this requires:

1. investments in conserving the remaining natural capital, such as groundwater stocks, soil, habitats for rare species;
2. encouraging the growth of natural capital by reducing our level of current exploitation, such as of non-renewable energy;
3. investments to relieve pressure on natural capital stocks by expanding cultivated natural capital, such as parks for inner-city recreation to relieve pressure on natural forests; and
4. Increasing the end-use efficiency of products, such as energy-efficient buildings, environmentally friendly urban transport.

1.7 Social Equity for Urban Sustainability

We, cities and towns, are aware that the poor are worst affected by environmental problems (such as noise and air pollution from traffic, lack of amenities, unhealthy housing, lack of open space) and are least able to solve them. Inequitable distribution of wealth both causes unsustainable behaviour and makes it harder to change. We intend to integrate people's basic social needs as well as healthcare, employment and housing programmes with environmental protection. We wish to learn from initial experiences of sustainable lifestyles, so that we can work towards improving the quality of citizens' lifestyles rather than simply maximising consumption.

We will try to create jobs which contribute to the sustainability of the community and thereby reduce unemployment. When seeking to attract or create jobs we will assess the effects of any business opportunity in terms of sustainability in order to encourage the creation of long-term jobs and long-life products in accordance with the principles of sustainability.

Continued

I.8 Sustainable Land-Use Patterns

We, cities & towns, recognise the importance of effective land-use and development planning policies by our local authorities which embrace the strategic environmental assessment of all plans. We should take advantage of the scope for providing efficient public transport and energy which higher densities offer, while maintaining the human scale of development. In both undertaking urban renewal programmes in inner urban areas and in planning new suburbs we seek a mix of functions so as to reduce the need for mobility. Notions of equitable regional interdependency should enable us to balance the flows between city and countryside and prevent cities from merely exploiting the resources of surrounding areas.

I.9 Sustainable Urban Mobility Patterns

We, cities & towns, shall strive to improve accessibility and sustain social welfare and urban lifestyles with less transport. We know that it is imperative for a sustainable city to reduce enforced mobility and stop promoting and supporting the unnecessary use of motorised vehicles. We shall give priority to ecologically sound means of transport (in particular walking, cycling, public transport) and make a combination of these means the centre of our planning efforts. Motorised individual means of urban transport ought to have the subsidiary function of facilitating access to local services and maintaining the economic activity of the city.

III.5. We shall seek to get our own house in order by implementing the principle of negotiating outward

Our community should not export its problems into the larger environment or to the future. The ethics of sustainable development require us to follow the principle of negotiating outward, which addresses the need to achieve a balance between local demand and supply in our city, and to the extent that this will not be possible, to enter into negotiations with our surrounding region, country, and continent about the share of opportunities, burdens, and responsibilities. We shall identify all options for the community to stop or reduce exporting problems and check if the conditions for using the larger environment and the future are equitable. We will undertake social, economic and eco-audits of our municipality in order to investigate its impact on the environment and future generations, and report on the results.

Accessed at: http://www.aalborgplus10.dk/media/
key_documents_2001_english_final_09-1-2003.doc March, 2007.

The emergence of the modern environmental movement in the late 1960s has not only led to important policy reforms that deal with pollution, it has also profoundly influenced urban design and development. Before the 1970s, ecology had little place in landscape design. But, by the 1980s, urban planners wrote of the importance of nature in the city, recognizing that it has powerful implications for how the city is designed, built and maintained. The landscape architect Anne Whiston Spirn called for re-integrating rather than exploiting nature in urban planning.[237] Many have referred to this as the "greening of the city". Green cities are already beginning to emerge, but there remains much to do to transform existing cities under green principles. There are numerous examples of ways that cities have begun to reconnect to the natural world within the urban landscape. Greening the city can involve tree-planting programs, brownfield redevelopment, heritage preservation, smart buildings, urban farms and urban forests, ecosystem restoration, bicycle-friendly cities, improved recycling programs, restricting the use of cars, expanding open spaces, and encouraging micro-development projects that help construct infrastructure in developing cities.

Green maps

The Green Map System (GMS) is a non-profit organization created in 1995. It aims to allow individuals and teams to create maps of their local community. These maps chart the natural and cultural environment and promote the rediscovery of the local environment.

All Green Maps are made locally by the people who live in the community. Each project is independently directed, funded and evaluated. Projects receive help from the Green Map network, and GMS's New York City office in getting started, sharing ideas and resources, promotion and advice. Registered Mapmakers receive a kit of Green Maps and a resource disk. The GMS organization invites people of all ages and backgrounds to create a unique Green Map of their home and to post it on the web site. Most maps are interactive and show hundreds of resources within the city.

Each local group registers their project, then spends many weeks or months researching their town's eco-cultural resources. They then design and publish their own regionally flavored Green Map. Even the smallest green site becomes important when it is literally put "on the map". Projects around the world – from neighborhoods and small towns to big cities – find that each effort is elevated by being part of the network.

Continued

Local map makers select from more than 125 icons to highlight green living resources. For example, there are icons to indicate organic farmers' markets, green business and green stores. There are icons that note the presence of toxic wastes, oil spills or noise pollution sources and ones that highlight wildlife habitats, dog parks and bird-watching sites. These icons are universal and link all the Green Maps together, helping green ideas and examples spread from city to city. Using the shared visual language of Green Map Icons, Green Maps cultivate citizen participation and community sustainability in hundreds of places around the world. The motto of GMS is *"When people see, they connect, then value and take care of a place."*

Source: Adapted from http://www.greenmap.com/

Green houses

Every house is a vortex of environmental consumption – of materials, energy and water. In the twentieth century, planners tended to view housing within a vast infrastructure grid of gas, electrical power, water supply and sewerage. It was easy to expand this basic pattern with more suburbs and new towns. Suburbs and sprawl ultimately encouraged forms of urban growth that were oblivious to their effects on the environment. In contrast, sustainability would localize the impacts of growth by insisting that as many resources as possible (water, energy and food) are sourced, processed and disposed of locally.[238]

A "green house" is one in which the building or rebuilding attempts to make the house self-sustaining. Table 12.2 lists the objectives of green housing. There are three main systems that contribute to sustainability 1) the waste system 2) the drinking water system and 3) the solar energy system. For example, one type of waste treatment system relies on a "biolytic filter" that takes sewage, good waste and other organic material. The filter consists of a concrete tank with several layers of filter beds that contain micro-organisms and worms that sift, sort, digest and treat the solid waste and waste water. The toilet, shower, bath, dishwater and sinks drain into a single sewer pipe that empties into the top filter bed of the tank. Once the micro-organisms and worms have done their job, any remaining water is pumped up to an ultraviolet lamp to kill any remaining bacteria. The water can be used in the garden.

Table 12.2 *Objectives of Green Housing and Green Communities*

- Minimize the use of resources (water, land, energy)
- Minimize production of waste
- Minimize use of toxic materials
- Integrate open space, green space into plan
- Minimize need for travel and maximize low-energy modes of transport (pedestrian, bike and public)
- Avoid privatized space, no gated communities (wasteful of land)
- Design public space for personal safety
- Insist on affordability and inclusiveness
- Produce some of the food consumed

It is also possible to use rainwater to satisfy water needs. Rainwater falling on the roof can be collected and filtered for drinking water, watering gardens and other fresh water needs. Solar systems can consist of solar panels of photovoltaic cells that produce electricity from sunlight. In some instances, there is enough energy power to sell back to the utility.

Green houses are also made using building materials from re-growth timber or materials produced by pollution-free manufacturing processes and no materials used that discharge toxic chemicals. For example, many use thick straw bales as insulation for walls. The use of concrete floors can also help keep the house warm in the winter, cooler in the summer. Many replace natural gas furnaces with stoves that burn corn kernels or wood pellets. Because corn kernels consume carbon dioxide as the plant grows, burning does not release new green-house gases.

One of the most important elements of a green house is the emphasis on design for the local climate and the orienting of the house so that main windows face south (in the northern hemisphere). This helps to maximize the use of the sun during the winter to provide light and warmth.

The impact of a green house can be considerable. For example, by using rainwater, a house can save 26,500 gallons of water that would have been con-sumed from a river or reservoir. By treating waste on the property the house keeps 26,500 gallons of sewage from flowing into a treatment plant (or being discharged without treatment during rain). Composting food scraps and other organic material can cut the solid waste that might otherwise be taken to a landfill. A house using solar energy reduces carbon dioxide emissions from electricity generated at a coal-fired power station by 8 tons.

Individuals and families that live in green houses are said to "live lightly on the grid" in that they have succeeded in creating homes that require only minimal energy from power plants and fossil fuels. The multiple benefits of a green house include avoiding high utility costs, electrical outages and combating climate change. The climate change connection is not inconsequential; the US EPA estimates that the average American house contributes more than twice as much greenhouse gas as the average automobile.

Green buildings

For many years architecture was rarely concerned with issues of sustainability. Sustainability often ranked below considerations of style and cost. More recently, however, green builders, architects and interior designers have developed designs for "green" buildings.

Buildings consume enormous quantities of the Earth's resources in both their construction and daily operation. The conceptual framework behind green buildings is the incorporation of features that support the conservation of the environment. Green buildings are often designed and oriented to minimize summer afternoon solar heat gain and optimize winter solar heat gain. Some may have solar energy as an alternative to fossil fuels. Many green builders select materials that do not have formaldehyde and have minimal or non-toxic properties to improve indoor air quality. Even interiors can incorporate materials and products that have high levels of renewability or reusability such as bamboo flooring or cork tiles.

Another example is in the installation of landscaping on the tops of building. Green roofs reduce energy costs and soak up rainwater. Green roofs are partially or completely covered with vegetation and soil, or a growing medium, planted over a waterproofing membrane. They were developed in Germany in the 1960s, and have since spread to many cities in Europe. Today, it is estimated that about 10 percent of all German roofs have been "greened".[239] Between 1989 and 1999, German roofing companies installed nearly 350 million square feet of green roofs and the rate is increasing. Although green roofs have become increasing popular in Europe, the adoption of green roofs in the US has been slower. Green roofs can impact the environment in the following ways: reduce carbon dioxide, reduce summer air conditioning needs, reduce winter heat demands, reduce stormwater runoff, provide songbird habitat, and remove nitrogen and other pollutants from rainfall. In addition, green roofs also reduce the urban heat island effect. Traditional building materials soak up the sun's radiation and reflect it back as heat, making cities at least 7° F hotter than surrounding areas. Figure 12.3 shows a green roof in Vienna, where rooftop temperatures on a hot day are

Figure 12.3 *Green Roof in Vienna. This building, called the Global Yard, is public housing that is comprised of 50% Vienna-born residents and 50% immigrants.*
Source: *Photo copyright Rob Crandall*

typically 25° F (14° C) cooler than they are on traditionally roofed buildings nearby.

In theory, suburban and urban housing developments can combine all of the elements of green housing and green buildings. While they contain a mix of activities from housing, workplaces, shops, cafes, sports facilities and schools, they also aim for zero net greenhouse emissions.[240]

Green spaces

In contemporary cities, one of the most prominent features in the marketing of residential areas is open space and natural features. Residential properties that line Central Park in New York, or even row houses that boast private gardens, are often the most desirable and sometimes the highest-priced properties. Trees, plants and other vegetation are not merely cosmetic embellishment; they are basic infrastructure that makes important contributions to the city aesthetically and ecologically.[241] For example, trees and shrubs shade the walls of houses and buildings, thereby reducing indoor temperatures and hence the need for air-conditioning. One study suggests that merely viewing natural landscapes can have a positive effect on people's sense of physical and psychological well-being.[242] Another study, published in *Science*, showed that hospital patients whose rooms had window views of trees and gardens had significantly higher rates of post-surgical recovery and shorter hospital stays than those whose rooms did not have views of vegetation.[243]

Sustainable approaches to green spaces can include restoration of creeks and waterways, planting street and garden trees, and the restoration of native plants and animals. Efforts to restore creeks not only help restore indigenous riparian plant species, but also help manage urban stormwater. Many creek and river restoration projects have the added element of providing residents and tourists with opportunities to experience these restored natural settings with the addition of new or improved walking trails. The greening of such small urban spaces is often initiated by urban community groups.

In New York City, the local grassroots organization Green Guerillas assists neighborhoods in creating community gardens and in developing garden coalitions to resist development of small urban open spaces. The organization dates back to 1973, when a Lower East Side artist, Liz Christy, organized her friends and neighbors to clean out a vacant lot on the corner of Bowery and Houston Streets. They created a vibrant community garden, thus establishing the modern community gardening movement in New York City. In other de-industrialized cities such as Pittsburgh and Detroit, abandoned, derelict lands have been given over to food growing by unemployed workers. There has been a revival of urban farming and agriculture.

Urban agriculture is not new to cities: it can be traced back to the earliest cities; even medieval cities in Europe grew crops within the walls. But it has be "rediscovered". Today local supplies of fruit and vegetables are part of sustainability. Community gardens that grow food for local residents reduce the energy required to transport food from hundreds or thousands of miles away. In many developing cities, the growing of food in and around cities contributes to food security and poverty alleviation. For example, in Dar es Salaam, Tanzania, one of the fastest growing cities, 67 percent of the families are engaged in urban farming, compared with 18 percent in 1970.[244] In Accra, Ghana, urban farmers supply 90 percent of the vegetables. In the shantytowns of Lima, Peru, inhabitants produce a variety of crops from sweet potatoes and artichokes to chicken, fish and pork. Many of these urban farmers recently immigrated from the Andes mountains, where agriculture has been a way of life, and their skills have been put to good use in the shanties.[245] Urban farming, and community gardens are examples of locally organized and locally managed green projects.

Increasingly, urban planners incorporate the principles of landscape and urban ecology into the siting and design of small patches of land, larger areas such as parks and even wider park systems. Many cities are currently planning to connect habitat patches through environmental corridors and networks of corridors. A good example of this can be found in Washington, DC. The city has two rivers, the Potomac to the west and the Anacostia River to the east. The Anacostia River has come to symbolize the physical and social divide – to the east are the "have nots". While there have been ongoing cleanup efforts with the Potomac, the Anacostia has been largely ignored and often treated as the capital's backyard. It became the location for unwanted land uses and neglectful land management practices. It was also the location for most of the city's public housing projects. Like many other urban rivers across the world, the Anacostia has long been viewed as a resource to exploit. By the 1990s, it had become a place to ignore or avoid, with much of its edge blocked by highways and many adjacent neighborhoods in decline. In 2000, the Anacostia was one of the ten dirtiest rivers in the US, located in one of the nation's most impoverished areas.

In 2000, the city announced the launch of the Anacostia Waterfront Initiative (AWI) to restore the river and revitalize the waterfront. The AWI is a partnership between the District of Columbia, the federal government and the community. This partnership was necessary because the city and the federal government own more than 95 percent of the riverfront land and 70 percent of the land around the waterfront. The plan is ambitious: to revitalize and reconnect 2,000 acres of long-neglected land along both sides of the eight-mile long Anacostia waterfront (see Figure 12.4 and Table 12.3).

There are many stakeholders. The Anacostia Watershed Society is the major non-governmental organization active in organizing citizens; its projects include monitoring water quality, providing fields trips to the river for schoolchildren and organizing quarterly cleanup days. Another citizen's group, The Friends of Sligo Creek, assigns individual stewards to one-mile sections of the creek. Beyond cleaning up garbage, these groups plant trees and fight against invasive species. To coordinate the vast array of government and neighborhood environmental

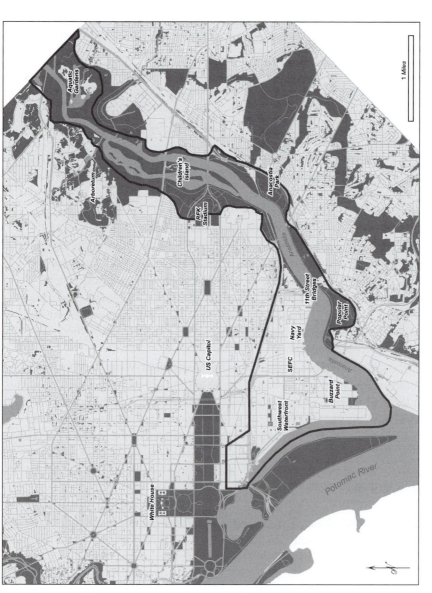

Figure 12.4 *Washington, DC's Anacostia Waterfront Initiative. The plan calls for the land inside the map boundaries to be connected via a system of trails and pathways and for public transportation to better link areas and neighborhoods along the waterfront.*

Source: Lisa Benton-Short

Table 12.3 *Anacostia Waterfront Plan Objectives*

- Connect 1,800 acres of park land
- Create 100 acres of new public parks
- Improve water quality
- Undertake wetland restoration
- Reconfigure transportation system to better serve neighborhoods
- Add 15,000–20,000 additional housing units in the area
- Add new mixed-use neighborhoods at Southwest Waterfront, Hill East Waterfront
- Encourage future museums and memorials
- Add 20 million square feet of commercial, retail and service-oriented space
- Create new cultural park celebrating the area's history and heritage
- Construct enhanced Boathouse Row and add two new boat launching sites
- Generate $ 1.5 billion over 20 years in additional tax revenues

Source: DC Office of Planning (2004) Anacostia Waterfront Initiative Framework Plan in pdf file. Accessed at: http://www.planning.dc.gov/planning/cwp/view,a,1285,q,582193,planningNav_GID,1708.asp

organizations, the city created a public–private partnership, the Anacostia Waterfront Corporation in 2003. The Anacostia Waterfront Initiative envisions a mixed-use network of residential space, offices, shopping, parks, trails, recreation areas and historic sites all around a green and clean river.

The Plan calls for the current much-neglected and fragmented parks to be integrated and connected into a twenty-mile-long river walk and trail. The parks will also connect back into neighborhoods and transportation networks thus ending the isolation of fragmented patches and instead integrating them into a unified waterfront park.

Although the plan waxes poetic, the reality is it will be very expensive to implement. Few of the publications or planning documents mention predicted costs, but estimates are in the billions. For example, it is estimated that it will cost $1.2 billion just to upgrade the Combined Sewer Systems to prevent CSO overflow releases from impacting on the river. In addition, there is the need for wetland restoration and pollution cleanup. Since this project is in its infancy and will not be completed for at least 10–15 years, the positive and negative impacts are difficult to judge. Yet the city is doing something it has never done before: plan for urban growth alongside ecological integrity.

Green transportation

In many cities, transportation is by car. However, the costs are substantial. Building roads, bridges and highways takes enormous sums of money. The rise of the car has also come at the expense of public transportation and a diversity of transportation choices. It contributes to congestion, road deaths and injuries. Car dependency also has ecological costs – it impacts on the atmosphere with greenhouse gases and local pollution, such as photochemical smog. The car, however, is fundamentally not a city vehicle, but it has dominated many urban transportation priorities.

A sustainable city will involve a transportation system in which there is a network of paths connecting people safely on foot, bicycle or public transportation. In city centers, space will be allocated to pedestrian zones, street cafes and markets instead of parking garages and congested city streets. When people do use cars, they will travel in fuel-efficient vehicles, or low-emission vehicles that emit very few pollutants. A system of green transportation prioritizes foot travel, then cycling, then public transport and finally private motor vehicles, a reverse of the customary order.[246] As one architect says, "a sustainable city is compact, polycentric, ecologically aware and based on walking".[247]

The commercial success of pedestrian precincts created in the historic centers of many European cities is an example of a re-prioritization of transport (see Figure 12.5). Because the car has been banned from these areas, shopping and working have been lively experiences. In the 1990s the city council in Birmingham, UK, decided that the historic nineteenth-century core of the city was not flourishing, in part because of a motorway that encircled the central business district. They demolished part of the motorway to allow full pedestrian access and today most of the city's central core is a walking precinct. In addition, the pedestrian areas are being linked to railway stations and major bus routes, to allow foot travel to move easily from pedestrian areas to other forms of transportation.

In cities such as Hong Kong, Vienna, Zurich, Curitiba and Singapore, new tram and bus systems are convenient and efficient. In cities such as Amsterdam there are expansive cycling networks that separate bicycles from automobiles, thus reducing the danger of accidents and providing fast and convenient pathways in and around the city. Bicycles have become a dominant form in the visual landscape (see Figure 12.6).

Investment in public transportation also addresses issues of social equity by providing transport for all those who do not have access to a car. Urban transportation

Figure 12.5 *Pedestrian area in Amsterdam. A convivial city depends upon restricting vehicular traffic in city centers.*
Source: Photo by John Rennie Short

that focuses on mass transit, walking and cycling means faster journeys, lower transportation costs and healthier people.

Green cities are not necessarily ones where people own fewer cars, but they are ones where people use (or need) cars less.[248] The idea of moving away from cars to greener forms of transportation seems sensible in principle, but has proven complex in practice. In the interim, programs that encourage car-sharing offer an alternative. In the US, many cities have FlexCar or ZipCar programs. Members pay a deposit and receive a key code that opens a variety of cars available all over the city (you can also locate them on the Internet sites). Car-sharing plans start at around $8 an hour or $60 a day for use of the car. Members share access to hundreds of vehicles, often within a five-minute walk of home or work. Members can reserve a car online or by phone and the hourly rate covers gas, insurance, monthly parking and maintenance. Studies have shown that each car-share vehicle can displace up to 15 privately owned cars off the road.

Many cities have employed programs to reduce automobiles by designating "congestion zones", areas where drivers must pay a congestion charge just to enter.

Figure 12.6 *Bicycles in Amsterdam.* Source: Photo by John Rennie Short

In London where drivers are charged £5 to enter the congestion zone, this strategy has reduced car traffic by as much as 25–30 percent and has increased the efficiency of buses and taxis.

For cities in the developing world, addressing transportation can raise a variety of other issues. Take the example of Bangkok, one of the world's most congested cities. In 2003, the Thai government announced a new measure for countering congestion in Bangkok: it was banning elephants from the city's streets. This ban affected some 250 elephants that had been wandering the city with their handlers, begging for food or selling trinkets. Often the elephants held up traffic, were hit by vehicles or fell into open drains.[249] Similarly in Hyderabad, India, notorious for air pollution and congestion, sacred cows roam the city streets, an added obstacle for rickshaws, taxis and cars to negotiate. To compound issues of congestion, many cities in the developing world have invested little in mass public transportation. Megacities need a comprehensive regional transport plan to adequately deal with the congestion. Another challenge for developing cities is that as incomes rise many residents are determined to catch up with the rich world in terms of automobile ownership. In 2002, China had a fleet of 15.5 million cars, one car for every 84 people (compared to one car for every 1.3 people in the US).[250] In cities such as Shanghai and Beijing, car sales have been increasing dramatically; the forecast is that China will see a 10-fold increase by 2020 to nearly 156 million vehicles.

As urban planners deal with greening transportation, they are challenged to remodel and regenerate existing cities and to reintegrate the diverse forms of transportation in urban life. The campaign for walkable cities offers the opportunity to reconnect urban residents in new pedestrianized inner city areas, reduce the need for the private car, and allow public transport and urban village models to prevail.

Greening the Olympic City

An example of how important the environment has become on the global stage was seen in the plans for Sydney to host the 2000 Summer Olympic Games. Sydney embodied a change in the hosting of the Summer Olympics Games toward a more environmentally sensitive event. It was not a new departure. The trend started after the Montreal Olympics in 1976, which marked the highpoint of massive new construction on essentially greenfield sites. In the wake of the enormous cost overruns and huge deficits of the Montreal Games, organizers started to look at alternatives. The Los Angeles Games of 1984 marked a shift toward refurbishment of existing facilitits rather than relying solely on the construction of brand-new facilities. Subsequent Games, as part of a way to reduce costs as well as to be more environmentally sensitive, used a mix of new build and refurbishment. But Sydney marked the beginning of an explicitly Green Games. The International Olympic Committee (IOC), the organization that selects which cities host the Games, influenced the shape of recent city bids by promoting the notion of a Green Games. Sydney won the competition to host the 2000 Summer Olympics, in part, because of its environmental initiatives.

It started with the main Olympic site, Home Bush Bay, which was an abandoned inner city area previously dominated by chemical manufacturers, brickworks and abattoirs. The area was badly contaminated. For the Games the site was cleaned up and the main stadium and other event sites were built with explicitly environmentally friendly designs. The Olympic Village, for example, was built on the site of a landfill and ammunition dump, with sustainable development in mind, utilizing solar power and recycled water. Over 50 percent of all the water used at the Games was recycled and most of the food and paper waste was composted. Sponsors and suppliers had to meet environmental standards in their business practices in order to receive contracts. Greenpeace was actively involved in both the initial bid and subsequent monitoring. When it was discovered that Coca Cola would employ 1,800 refrigerators at the games that used ozone-depleting refrigerants, Greenpeace activists protested. Coca Cola responded by announcing a phase-out of greenhouse gas refrigerants. The global stage of the Summer Olympics provides companies and corporations with an opportunity to position themselves in a global market. As sophisticated consumers become more demanding, companies can increase market visibility and market share by aligning with Green Games.

The Games also provide a global showcase for green city initiatives. Other cities can learn from host cities while designers and planner can see how green initiatives work out in practice.

How green was the Sydney Olympics? A contaminated urban site was cleaned up, the Olympics village became a new more sustainable suburb and new environmental practices were established and improved upon. The experience of Sydney has allowed the development of criteria and benchmarks for a Green Olympics. They include

- Use or adapt existing buildings
- Environmentally friendly building and design
- Minimize adverse impacts
- Protect local environments
- Build on previously used industrial and commercial sites so Olympics becomes part of greening the city
- Make sites accessible by clean public transport

While Sydney did not meet all of the criteria, the 2000 Games provided an important step toward a Green Games and ultimately toward Greener Cities. Beijing, which will host the 2008 Summer Games, is one of the world's most populous and polluted cities. China's plans for the host city include high-speed railway, new subway lines, an energy efficient airport terminal, planting millions of trees, improving sewage treatment and cleaning up its pollution. Plans called for dismantling or relocating some 190 steel, cement, chemical, paper and other factories; old high-emissions buses and taxis have been replaced with low-emissions vehicles. Beijing's municipal government pledged $13 billion for environmental cleanup and protection. Although the city has made tremendous progress, air pollution remains above acceptable levels. By 2007, the Chinese had spent $40 billion, but were finding it difficult to undo the environmental damage and turn Beijing into a green showcase. Despite these shortcomings, it is clear that there is now an international expectation that an Olympic host city can improve environmental quality and address urban sustainability.[251]

Curitiba: a Green City in the developing world

Much of our discussion of cities in the developing world has stressed "brown" issues such as poor sanitation, water quality, air pollution and housing problems. However, all is not doom and gloom. Curitiba, Brazil, is located near the coastal mountain range in the southern part of the country. Like many of Brazil's cities, it developed rapidly in the second half of the twentieth century, growing from 500,000 inhabitants in 1965 to 2.1 million by 1990. Such rapid growth brought the typical urban problems of unemployment, slums, automotive gridlock, pollution and environmental deterioration. Yet despite this, Curitiba is often cited as a model "Green City". In the late 1960s and 1970s, Curitiba's political elites, led by its three-term mayor, Jaime Lerner, encouraged urban planners to think imaginatively, integrating social and ecological concerns. The planning process created innovative and modest solutions in public transportation, recycling, garbage collection and green space expansion, many of them models of sustainable urban planning.

Curitiba was faced with an inefficient public transportation system and an increase in the number of private automobiles. Initially, planners leaned toward the development of a subway system, at the cost of some $60–$70 million per kilometer. Instead, they turned to modifying the bus system, at $200,000 per kilometer, or one percent of the cost of the subway. To meet the growing needs for transportation, and to curtail the use of private automobiles, planners focused on encouraging Curitiba's physical expansion along linear axes, each with a central road that had a dedicated lane for express buses. Some called this a "surface subway". The aim was to reduce congestion while returning the central area to the pedestrian. The use of a series of express buses was far cheaper than subways or light railways and highlights a more practical and affordable solution to public transportation in the developing world. The city was able to keep public transportation affordable; average low-income residents spend only 10 percent of their income on transport. In addition, modifications to the buses and boarding tubes have made the system very efficient. For example, along the bus routes are clear tubular structures that are level with buses, providing quick boarding and exiting and easier access for the handicapped. The buses themselves have five lateral doors along the side and can load three times as many passengers per hour as traditional buses. Although the city has more than 500,000 private cars (more cars per capita than most Brazilian cities), most residents use the bus system. In fact, the public transportation system is used by more than 1.3 million passengers each day, nearly two-thirds of the urban population.[252] One effect has been improved air quality in the city.

An innovative solution to garbage has been the city's recycling program. Introduced in 1989, the recycling program encourages city residents to separate organic and inorganic garbage. A city-wide environmental education program helped to reinforce the benefits of recycling. Over 70 percent of the community participates in the separation program, one of the highest rates for

Figure 12.7 *A local development organization. This group, GROOTS, in Kenya is a network of women that seeks to find ways for women to directly participate in development decision making, planning and implementation of issues that affect them. This is an example of sustainable development at a grassroots level. (For more information, go to http://www.groots. org/members/keny a.htm)* **Source: Photo by David Rain**

any city in the world. A second program is the "Purchase Garbage" program, aimed primarily at the shanty settlements in Curitiba. In the favelas, there is no organized garbage collection. To help deal with potential health problems, the program encouraged favela residents to collect garbage in return for bus tickets and groceries. This program has led to a decrease in city litter while helping to feed the urban poor. At one point, there were 22,000 families involved in the Purchase Garbage program, feeding approximately 100,000 people while collecting some 400 tons of trash. In 1990 Curitiba received an award from the United Nations Environment Programme for these two successful waste management programs.

Small scale, modest changes have also had positive impacts on the city. In the 1980s, Mayor Lerner distributed 1.5 million tree seedlings for neighborhoods to plant and care for, adding to green space improvement. Once planted, a tree cannot be cut without a permit, and for every tree that is felled, two must be planted. Green space has also expanded with the construction of 17 new parks, 90 miles of bike paths and trees everywhere. Many of the parks were created when planners diverted water from the lowlands (which were prone to flooding) into lakes that were the centerpieces of the new parks. Integrating flood control with park space development is an example of "design with nature". In addition, developers receive tax breaks if their projects include green areas. As a result, the city-wide ratio of open space to inhabitant has increased from 0.5 square meters to 52 square meters, which means that Curitiba has one of the highest averages of green space per urban inhabitant anywhere in the world.[253]

Under Lerner's administration more than 30 percent of the city's budget went to providing health care (more than three times other Brazilian cities).[254] Infant mortality rate dropped more than 60 percent over a 20-year period. The city has also invested substantially in improving primary education, retraining teachers and renovating public school buildings. Mayor Lerner convinced private firms to build more than 30 neighborhood day-care centers, run by the city. This partnership works because the firms' employees can use the day-care centers, a service that would otherwise be costly for companies to provide on their own.

The case of Curitiba provides valuable lessons not only to other developing cities, but to large cities everywhere. Curitiba rejected conventional wisdom that emphasized technologically sophisticated solutions to urban problems. Creative and labor-intensive ideas – especially where unemployment is a problem – can substitute for capital-intensive, high-technological solutions.[255] Planners have learned that the solutions to urban woes are not specific and isolated, but rather interconnected. Improving public transportation or refuse collection is inextricably bound up with issues such as poverty, crime, education and public health. The lesson is crucial: cities have the capacity to transform their environmental problems into creative and innovative solutions. Moreover, as the case of Curitiba shows, the integration of social justice and social services and environmental quality is central to sustainable urban development.

Conclusions

Strategies to achieve sustainability include planning for environmental issues related to air quality, energy efficiency, integrated water resources management, waste stream management and others. Through the efficient use of natural resources and switching from fossil to solar energy in its various forms (renewables), more livable and stable environmental conditions can be provided for the community and its economy in the longer term. A green city is better prepared for future global environmental resource conditions.

One challenge to achieving urban sustainability is the fragmentation of urban planning. Urban planners must abandon the current compartmentalization of planning specialties (i.e. housing, transportation, land use, etc.). Instead planning efforts need to be more holistically integrated. As we explored in Chapters 3 and 4 many nineteenth century planners such as Ebenezer Howard, Frederick Law Olmsted, Patrick Geddes and Daniel Burnham created urban designs that factored in multiple elements such as natural resources, parks, transportation systems, regional planning and social equity.

Another challenge is political. Urban sustainability is increasingly tied into issues of governance and political control.[256] While and colleagues propose the notion of the "urban sustainability fix" to describe the selective incorporation of ecological perspectives into city government. With reference to English cities, they show how entrepreneurial urban regimes have incorporated the green agenda, as in the case of Leeds, while others have sought to insulate themselves from dissent, as in the case of Manchester. In effect, their work shows the greening of the urban growth machine.[257]

Can cities effectively deal with growth and urban sprawl? Can cities achieve true sustainability? Are current levels of and projected levels of urbanization compatible with urban sustainability? These questions are among the most dramatic and for the time being, unanswered, questions facing us in the twenty-first century. Given that cities are the principal habitats of human life, the task at hand is a daunting but unavoidable one.

Guide to Further Reading

Adams, W. M. (2001) *Green Development: Environment and Sustainability in the South*, London: Routledge. [see http://www.geog.cam.ac.uk/people/adams/]

Barrow, C.J. (2006). *Environmental Management for Sustainable Development*, New York: Routledge.

Beatley, T. (2000) *Green Urbanism: Learning from European Cities*, Washington, DC: Island Press.

Breen, A. and Rigby, D. (1997) *Waterfronts: Cities Reclaim Their Edge*, Washington, DC: Waterfront Press.

Bullard, R. (2007) *Growing Smarter*, Cambridge, MA: MIT Press.

Curwell, S., Deakin, M. and Symes, M. (eds) (2005) *Sustainable Urban Development Volume 1: The Protocols and Environmental Assessment Methods*, London: Routledge.

Evans, P. (2002) *Livable Cities? Urban Struggles for Livelihood and Sustainability*, Los Angeles and Berkeley: University of California Press.

Girardet, H. (2004) *Cities, People, Planet: Livable Cities for a Sustainable World*, Chichester, England: Wiley Academy Press.

Fishman, R. (1977) *Urban Utopias in the Twentieth Century*, New York: Basic Books.

Irazábal, C. (2005) *City Making and Urban Governance in the Americas: Curitiba and Portland*, Burlington: Ashgate.

Kaplan, R., Kaplan, S. and Ryan, R. (1998) *With People in Mind: Design and Management of Everyday Nature*, Covelo, CA: Island Press.

Low, N., Gleeson, B., Green, R. and Radović, D. (2005) *The Green City: Sustainable Homes, Sustainable Suburbs*, Oxford: Routledge.

Low, N. and Gleeson, B. (eds) (2003) *Making Urban Transport Sustainable*, Basingstoke: Palgrave.

Naess, P. (2006) *Urban Structure Matters*, London: Routledge.

Moore, S. (2007). *Alternative Routes to the Sustainable City: Austin, Curitiba, and Frankfurt,* Lanham: Lexington Books.

Orr, D. (2006) *Design on the Edge*, Cambridge, MA: MIT Press.

Platt, R. (ed.) (2007) *The Humane Metropolis: People and Nature in the 21st Century*, Boston: University of Massachusetts Press.

Pezzoli, K. (2000) *Human Settlements and Planning for Ecological Sustainability: The Case of Mexico City*, Cambridge, MA: MIT Press.

Poole, B. (ed.) (2006) *Green Design,* New York: Mark Batty Publisher.

Pugh, C. (ed.) (1996) *Sustainability, the Environment, and Urbanization*, London: Earthscan.

Purvis, M. and Grainger, A. (2004) *Exploring Sustainable Development: Geographical Perspectives,* Sterling, VA: Earthscan.

Register, R. (2006) *Ecocities*, Gabriola Island, BC: New Society.

Robbins, E. and El-Khoury, R. (eds) (2004) *Shaping the City: Studies in History, Theory and Urban Design*, New York: Routledge.

Sorensen, A., Marcotullio, P., and Grant, J. (eds) (2004) *Towards Sustainable Cities: East Asian, North American, and European Perspectives on Managing Urban Regions*, Burlington, VT: Ashgate.

Wheeler, S. (2004) *Planning for Sustainability: Creating Livable, Equitable and Ecological Communities*, London: Routledge

For more on the Green Guerillas of New York City, visit: http://www.greenguerillas.org

For more information about green roofs, visit: http://www.greenroofs.org/

Postscript

Having covered a wide variety of material and a myriad of urban environmental issues let us conclude with a summary of the three underlying themes that we first raised in the introductory chapter: urban environments in historical context; issues in urban–nature relations; realigning urban–nature relations.

The city and nature: historical trends

The most striking historical trend has been for urbanization to increase in size and scope. We are moving toward an urban future. As more people live and work in cities the quality of the urban environment is of major importance. As we noted in Chapter 1, environmental issues are not incidental to the urban condition, they are central.

In Chapter 2 we examined how increasing urbanization created perceived environmental problems that in turn promoted new systems of regulation and infrastructural modifications that in turn generated changes to the city–environmental relationship.

A major theme in the history of city–nature relations is the rise of the industrial city. As Chapter 3 contends, the rise of the industrial city involved a marked contamination of water, land and air, subsequent policy reforms and new forms of urban design. The industrial city is not only a "city of dreadful night", it also is the impetus for realigning urban–nature relations through movements such as the urban parks movement and garden cities. Coping with the problems of urban environments during industrialization has been an important stimulus for new knowledge and the creation of new policies. Science and governments were employed to deal with the issues of providing safe urban environments, although the provision and the definition of "safe" are, of course, mediated through the prism of social and economic power and technological ability.

In Chapter 4 we identified five themes in the current wave of urban transformation; the impact of giant urban regions and megacities; postindustrial cities and brownfields, urban sprawl, new industrial spaces and shantytowns. Many cities are experiencing combinations of these spaces simultaneously, an additional challenge to urban policy and environmental regulation.

Urban environmental issues

We looked at contemporary urban environmental issues from two main perspectives. On the one hand, we discussed various ways to consider the urban–nature dynamic. Chapter 5 argued that the occupancy of specific sites creates constraints and opportunities and the setting for the production of urban images. In certain sites cities are more vulnerable to disasters. We noted, in Chapter 6, that there is no such thing as a "natural disaster". We preferred the term environmental hazard/ disaster. Disasters are rarely natural and their effects are unevenly distributed through the hierarchies of social and economic difference. The emergence of an urban ecology that we documented in Chapter 7 is still at an early stage but its future development will allow us a more rigorous look at the cities as social-biophysical complexes.

On the other hand, we also considered issues of pollution as another angle on understanding urban environmental issues. In Chapters 8, 9 and 10 we looked at water, air and garbage, respectively. In some cases we could point to improvements and advances especially in the developed world. In the case of the developing world it was a more mixed picture. Most cities remain confronted with significant issues of water, air and land pollution.

Realigning urban–nature relations

The realignment of urban–nature relations has to begin with a clear view of the connections between society and nature. In Chapter 11 we explored how issues of class, race and gender interconnect with environmental issues and have become increasingly important areas of research and documentation. There remains much to do to fully understand the complex interconnections between race, class and social and environmental justice.

Another fundamental realignment of urban–nature relationships are those discourses that seek to reconstruct the twentieth-century metanarrative, which sought to dominate and control nature through a policy of uncontrolled growth. Slow Growth, New Urbanism, Smart Growth, historic preservation and urban sustainability are just some of the examples of these new urban environmental discourses. These discourses, explored in Chapter 12, attempt to redesign, recreate and rethink cities within a larger framework of livability and sustainability. The issue of sustainable urban development not only reconnects cities to their local and regional ecologies but also raises issues of participatory democracy. Sustainability is not simply an ecological issue, it is a political and economic issue.

Another angle from which to understand the urban–nature nexus is global climate change. Early debates about climate change tended to focus on its potential impact on wilderness areas such as the polar ice caps and on wildlife, plant life and biodiversity. But what is becoming apparent is that global climate change will have major urban impacts. Increasing temperatures for example, will warm the urban heat island. More hot days will result in older, poorer people in affected cities being more at risk of heat exhaustion. Rising sea levels will make many coastal cities more vulnerable to flooding. Climate change will increase the intensity of hurricanes and typhoons with dramatic effects on cities subject to such hazards. Warmer temperatures also create new pathways for disease. Mosquitoes can now survive at higher altitudes making some cities, previously too high, more vulnerable to malaria outbreaks. Warmer wetter climates are making residents of some cities more vulnerable to new viruses, such as West Nile. The impact of all the global climate change in cities will weigh heaviest on the poor, the sick, the old and the marginal.

Around the world the effects of global climate change have the potential and actual ability to realign human-environment relations. Global climate change, in part produced by the nature of urbanization with its heavy energy demands is, in turn, restructuring the relationship of many cities to the physical environment. Assuming no significant reductions in the emissions of greenhouse gases, the models suggest a one degree centigrade increase by the 2020s with increasing morbidity and mortality from heat waves, floods and droughts. By the 2050s, a two-degree increase is forecast with millions facing increased risk of flooding. By the 2080s the projected four-degree increase will create global economic losses averaging up to 5 percent of GDP. The projections assume no change in emissions. If emissions increase the affects will occur more rapidly and probably more severely. If emissions decrease significantly and quickly the worst affects may be minimized if not averted.

Cities and global climate change

Climate disruption is an urgent threat to the environmental and economic health of our communities. Many cities already have strong local policies and programs in place to reduce global warming pollution, but more action is needed at the local, state, and federal levels to meet the challenge. On February 16, 2005, the Kyoto Protocol, the international agreement to address climate disruption, became law for the 141 countries that have ratified it to date. In response, mayors in cities such as Seattle and London called for more radical local action to reduce global warming pollution.

In March 2005, Seattle Mayor Greg Nickles and nine other US mayors representing more than 3 million Americans, joined together to invite cities from across the country to take additional actions to significantly reduce global warming pollution. Three months later in June, the Mayors' Climate Protection Agreement was passed unanimously by the U.S. Conference of Mayors. By April 2007, 453 mayors in the US had signed the agreement.

Under the Agreement, participating cities commit to take following three actions:

- Strive to meet or beat the Kyoto Protocol targets in their own communities, through actions ranging from anti-sprawl land-use policies to urban forest restoration projects to public information campaigns;
- Urge their state governments, and the federal government, to enact policies and programs to meet or beat the greenhouse gas emission reduction target suggested for the United States in the Kyoto Protocol – 7% reduction from 1990 levels by 2012; and
- Urge the US Congress to pass the bipartisan greenhouse gas reduction legislation, which would establish a national emission trading system.

In October 2005, representatives of some 20 cities met in London at the World Cities Leadership Climate Change Summit, which was organized by the Mayor of London, Ken Livingstone. This conference launched an international collaboration on the pressing issue of climate change. The conference brought together those city leaders who are taking the most productive and radical steps to adapt to and mitigate climate change, to share ideas and provide leadership to the rest of the world. One result of the conference was the C40 Group. The C40 group is a group of the world's largest cities committed to tackling climate change to reduce carbon emissions and increase energy efficiency in large cities across

Continued

the world. C40 members include: Bangkok, Berlin, Bogotá, Buenos Aires, Cairo, Caracas, Chicago, Delhi NCT, Dhaka, Houston, Istanbul, Jakarta, Johannesburg, London, Los Angeles, Madrid, Melbourne, Mexico City, Moscow, New York, Paris, Philadelphia, Rome, São Paulo, Seoul, Tokyo and Toronto. Smaller, affiliated cities include: Austin, Barcelona, Copenhagen, Curitiba, Heidelberg, New Orleans, Portland, Rotterdam, Salt Lake City, San Francisco, Seattle and Stockholm.

London has adopted its own Climate Change Action Plan that outlines numerous measures that aim to save London 20 million tons of carbon by the year 2025.

Sources: adapted from http://www.london.gov.uk/mayor/environment/climate-change/ And http://www.seattle.gov/mayor/climate/

Let us end the book with a recurring theme: the city is part of the environment and the urban environment is a social and political construct. Karl Marx once noted that people make their own history but not in circumstances of their own making. We make our urban environments and the difference between history and geography is that, collectively, we can probably make our own circumstances. But we have to be aware of the fragility of these circumstances. As the cosmonaut Pavel Popovich reminds us, from the unique perspective of someone who has looked down on the earth,

> It seems to me that even the wisest of philosophers of the Renaissance, or the most daring minds from the past could not estimate the real size of our planet. Earlier it seemed immeasurably great, almost infinite. Only after the middle of this [20th] century did many, having gone up above the Earth into space, see with surprise and disbelief just how small the Earth really is. Some saw it as an island in the limitless ocean of creation. Some compared it to a spaceship with a crew numbering more than 6 billion. I was terrified by its fragile appearance.

Notes

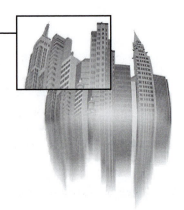

Chapter 1 The city and nature

1. Klineneberg, E. (2002) *Heat Wave: A Social Autopsy of Disaster in Chicago,* Chicago: University of Chicago Press.

2. Keil, R. (2003) "Urban political ecology", *Urban Geography* 24: 723–738.

3. Solecki, W. and Rosenzweig, C. (2004) "Biodiversity, biosphere reserves and the big apple, a study of the New York metropolitan region", *Annals of New York Academy of Science* 1023: 105–128.

4. Douglas, I. (1981) "The city as an ecosystem", *Progress in Physical Geography* 5: 315–367.

5. The Baltimore Ecosystem Study at http://www.beslter.org/ and the Central Arizona–Phoenix Study at http://caplter.asu.edu/

6. Panayotou, T. (2001) "Environmental sustainability and services in developing global city–regions", in Scott, A, J. (ed.) *Global City–Regions,* Oxford: Oxford University Press.

7. Evans, P. (ed.) (2002) *Livable Cities: Urban Struggles for Livelihood and Sustainability,* Berkeley and Los Angeles: University of California Press.

8. Houghton, M. (1999) "Environmental justice and the sustainable city", in Satterthwaite, D. (ed.), *Sustainable Cities,* London: Earthscan.

9. Hough, M. (2004, 2nd edn) *Cities and Natural Process: A Basis for Sustainability,* London: Routledge.

10. Gandy, M. (2002) *Concrete and Clay: Reworking Nature in New York City,* Cambridge MA: MIT Press.

11. Swyngedouw, E. (2004) *Social Power and the Urbanization of Water,* Oxford: Oxford University Press.

12. Cadenasso, M. L., Pickett, S. T. A. and Grove, M. J. (2006) "Integrative approaches to investigating human–natural systems; the Baltimore ecosystem study", *Natures Sciences Societies* 14: 4–14.

13. Keys, E., Wentz, E. A. and Redman, C. L. (2007) "The spatial structure of land use from 1970–2000 in the Phoenix, Arizona Metropolitan Area", *The Professional Geographer* 59: 131–147.

Chapter 2 Environmental issues in cities

14. Mumford, L. (1989) *The City in History,* San Diego and New York: Harvest Books, p. 71.

15. Pfeiffer, J. (1980) "The mysterious rise and decline of Monte Albán", *Smithsonian,* February.

16. Hardoy, J. (1973) *Pre–Columbian Cities.* Translated by Judith Thorne, New York: Walker, p. 33.

17. Mumford, L. (1989) p. 75.

18. Mumford, L. (1989) p. 75–77.

19. Markham, A. (1994) *A Brief History of Pollution,* New York: St Martin's Press.

20. It should be noted that not all cities built walls; some cities were fortified only when political leaders felt they were on threatened frontiers.

21. Lazzaro, C. (1990) *Italian Renaissance Garden: From the Conventions of Planting, Design, and Ornament to the Grand Gardens of Sixteenth–Century Central Italy,* New Haven: Yale University Press.

22. Merchant, C. (1980) *The Death of Nature: Women, Ecology and the Scientific Revolution,* London: Wildwood House.

23. Merchant, C. (1980) p. xviii.

24. Miller, I. (2002) *Washington in Maps 1606–2000,* New York: Rizzoli International Publications, p. 20.

25. Miller, I. (2002), p. 18.

26. Bowling, K. (1991) *The Creation of Washington, DC: The Idea and Location of the American Capital,* Fairfax, VA: George Mason University Press, p. 224.

27. Scott, P. (2002) "This vast empire: the iconography of the Mall, 1791–1848", in R. Longstreth (ed.), *The Mall in Washington, 1791–1991,* New Haven: Yale University Press. p. 39.

28. Smith, H. (1967) *Washington, D.C: the Story of Our Nation's Capital,* New York: Random House, p. 13.

29. Scott, P. (2002) p. 43.

30. Karlen, A. (1995) *Man and Microbes: Disease and Plagues In History and Modern Times,* New York: G.P. Putnam, p. 52.

31. Karlen, A. (1995).

32. Quoted in Sjoberg, G. (1960) *The Preindustrial City,* New York: The Free Press, p. 93.

33. Boccaccio, G. (1972) *The Decameron.* Translated by G.H. McWilliam, New York: The Penguin Classics.

34. Bell, G.W. (1994) *The Great Plague of London,* London: Bracken Books.

35. Swanson, M. (1995) "The sanitation syndrome: bubonic plague and urban native policy in the Cape Colony, 1900–09", in William Beinart and Saul Dubow (eds) *Segregation and Apartheid in Twentieth-Century South Africa,* New York: Routledge, p. 25–42.

36. Swanson, M. (1995) p. 26.

37. Karlen, A. (1995) p. 62.

38. Herlihy, D. (1970) *The History of Feudalism,* New York: Harper and Row, p. 270–271.

39. Quoted in McKay, J., Hill, B. and Buckler, J. (1995,5th edn) *A History of Western Society,* Boston: Houghton Mifflin, p. 347.

Chapter 3 The industrial city

40. Engles, F. (1973) *The Condition of the Working-Class in England,* Moscow: Progress Publishers, pp. 86-89, and 92.

41. Mumford M. (1989) *The City in History,* San Diego: Harvest Books, p. 446.

42. Melosi M. (1980) "Environmental crisis in the city: the relationship between industrialization and urban pollution", in Martin Melosi (ed.), *Pollution and Reform in American Cities, 1879–1930,* Austin: University of Texas Press, p. 6.

43. Karlen, A. (1995) *Man and Microbes: Disease and Plagues In History and Modern Times,* New York: G.P. Putnam, p.51.

44. Dennis, R. (1994) *English Industrial Cities of the 19th Century,* Cambridge: Cambridge University Press; and Mumford, L. (1989) *The City in History,* San Diego: Harvest Books, pp. 467–468.

45. Melosi, M. (1980) p. 7.

46. Markham, A. (1994) *A Brief History of Pollution,* New York: St Martin's Press, p. 20.

47. As quoted in Markham, A. (1994) p. 16.

48. Markham, A. (1994) p. 16.

49. Tarr, J. (1996) *The Search for the Ultimate Sink: Urban Pollution in Historical Perspective,* Akron, OH: University of Akron Press, pp. 8–9.

50. Melosi, M. (1980) p. 14.

51. Tarr, J. (1996) p. 324.

52. Tarr, J. (1996) p. 326.

53. Tarr, J. (1996) p. 14.

54. Tarr, J. (1996) pp. 14–15.

55. Markham, A. (1994) p. 16–20.

56. Robson, B. T. (1996) "The saviour city: beneficial effects of urbanization in England and Wales", in Douglas, Hussert and Robinson (eds), *Companion Encyclopedia of Geography,* New York: Routledge. pp. 300–301.

57. Galishoff, S. (1980) "Triumph and failure: the American response to the urban water supply problem, 1860–1923", in Martin V. Melosi (ed.) *Pollution and Reform in American Cities, 1870–1930,* Austin: University of Texas Press, pp. 35–58.

58. Boyer, M. C. (1986) *Dreaming The Rational City: The Myth of American City Planning,* Cambridge MA: MIT Press, p. 17.

59. Galishoff, S. (1980) p. 36.

60. Galishoff, S. (1980) p. 48.

61. Kaika, M. (2006) "Dams as symbols of modernization: the urbanization of nature between geographical imagination and materiality", in *Annals of the Association of American Geographers,* 96(2): 276–301.

62. Tarr, J. (1996) pp. 344–345.

63. Tarr, J. (1996) pp. 344–345.

64. Tarr, J. (1996) pp.12–13.

65. Melosi, M. (1980) p. 107.

66. Melosi, M. (1980) p. 113.

67. Grinder, R.D. (1980) "The battle for clean air: the smoke problem in post–civil war America", in Martin V. Melosi (ed.) *Pollution and Reform in American Cities, 1870–1930,* Austin: University of Texas Press.

68. Tarr, J. (1996) p. 16.

69. In Europe, royal parks and gardens provided only limited pubic access, usually on holidays, but there were exceptions. The Tiergarten in Berlin was opened to the public for "pleasure strolling" in 1649 and there were other public promenades.

70. Schulyer, D. (1986) *The New Urban Landscape: the Redefinition of City Form in Nineteenth Century America,* Baltimore: The Johns Hopkins University Press, p. 59.

71. This contrasts with existing urban parks in many European cities. There, existing gardens and parks were molded by earlier historical conditions and carried the imprint of different class–cultural relations. Hyde Park in London was open to the public (for a fee) in 1652 and in the 1830s the palace and gardens of Luxembourg and Versailles became public.

72. Peterson, J. (1996) "Frederick Law Olmsted Sr. and Frederick Law Olmsted Jr.: the visionary and the professional", in Mary Corbin Sies and Christopher Silver (eds) *Planning the Twentieth Century American City,* Baltimore: The John Hopkins University Press, pp. 37–54.

73. Gopnik, A. (1997) "Olmsted's Trip" in the *New Yorker Magazine,* March 31, p. 101.

74. Quoted in Cranz, G. (1982) *The Politics of Park Design: A History of Urban Parks in America,* Cambridge: The MIT Press, p. 29.

75. Cranz, G. (1982) p. 34.

76. Olmsted, F.L. (1996) "Public parks and the enlargement of towns", in Richard T. LeGates and Frederic Stout (eds) *The City Reader,* London and New York: Routledge, p. 339.

77. Olmsted, F.L. (1996) p. 343.

78. Cranz, G. (1982) *The Politics of Park Design: A History of Urban Parks in America,* Cambridge: The MIT Press, p. 41.

79. Schulyer, D. (1986) *The New Urban Landscape: the Redefinition of City Form in Nineteenth Century America,* Baltimore: The Johns Hopkins University Press, pp. 64–65.

80. Schulyer, D. (1986) p. 78.

81. Fishman, R. (2003) "Urban utopias: Ebenezer Howard, Frank Lloyd Wright and Le Corbusier", in Scott Campbell and Susan Fainstein (eds) *Readings in Planning Theory, Second Edition,* Oxford: Blackwell, p. 22.

82. Howard, E. (1996) "Author's introduction from garden cities of to–morrow", in Richard T. LeGates and Frederic Stout (eds) *The City Reader,* London and New York: Routledge, pp. 346–353.

83. Freestone, R. (1989) *Model Communities: The Garden City Movement in Australia,* Melbourne: Melbourne University Press.

Chapter 4 Contemporary urbanization and environmental dynamics

84. This paragraph draws heavily on the Introduction in Hall, T., Hubbard, P. and Short, J. R. (eds) (2007) *The Sage Urban Compendium,* London: Sage.

85. United Nations Centre for Human Settlements, (2001) *Cities in a Globalizing World: Global Report on Human Settlements,* 2001, London and Sterling, VA: Earthscan, p. 3; Short, J. R. (2004) *Global Metropolitan,* London: Routledge.

86. Short, J. R. (2004) *Global Metropolitan,* London: Routledge.

87. Taylor, P. (2004) *World city networks,* London: Routledge.

88. Short, J. R. (2007) *Liquid City,* Washington, DC: Resources For The Future.

89. Castells Manuel. (1998) "Why the megacities focus? Megacities in the New World disorder", *The Mega–Cities project Publication MCP–108,* p. 1.

90. Haiqing, L. (2003) "Management of coastal mega-cities – a new challenge in the 21st century", *Marine Policy* 27: pp. 333–337.

91. Phienwej, M. and Nutalaya, P. (2005) "Subsidence and flooding in Bangkok", in Gupta, A. (ed.), *The Physical Geography of Southeast Asia,* London: Oxford University, pp. 358–378.

92. Haiqing, L. (2003).

93. The Blacksmith Institute (2006) *The Worlds' Worst Polluted Places: The Top Ten,* New York: The Blacksmith Institute. Report available online at www.blacksmithinstitute.org

94. Gordon, D. (1997) "Managing the changing political environment in urban waterfront development", *Urban Studies* 34(1): 61–83.

95. US EPA. (2007) Brownfields cleanup and redevelopment home page: http://www.epa.gov/swerosps/bf/

96. Alberini, A., Heberle, L., Meyer, P. and Wernstedt, K. (2004) "The brownfields phenomenon: much ado about something or the timing of the shrewd?" *Working Paper, Center for Environmental Policy and Management,* University of Louisville. Obtained at htt:/cepm.louisville.edu.

97. Howland, M. (2003) "Private initiative and public responsibility for the redevelopment of industrial brownfields: three Baltimore case studies", *Economic Development Quarterly,* 17(4): 367–381.

98. Grimski, D. and Ferber, U. (2001) "Urban brownfields in Europe", *Land Contamination & Reclamation,* 9(1): 143–148.

99. De Sousa, C. (2003) "Turning brownfields into green space in the city of Toronto", *Landscape and Urban Planning,* 62: 181–198.

100. De Sousa, C. (2003) p. 181–198.

101. Deutz, P. (2004) "Eco–industrial development and economic development: industrial ecology or place promotion", *Business Strategy and the Environment* 13: 347–362.

102. Kunstler, H. J. (1993) *The Geography of Nowhere,* New York: Touchstone; Duany, A., Plater–Zyberk, E. and Speck, J. (2000) *Suburban Nation: The Rise of Sprawl and The Decline of the American Dream.* New York, North Point Press; Putnam, R. D., (2000) *Bowling Alone: The Collapse and Revival of American Community,* New York: Simon and Schuster.

103. Bruegmann. R. (2005) *Sprawl: A Compact History,* Chicago: University of Chicago Press.

104. Wolman, H., Galster, G., Hanson, R., Ratcliffe, M., Furdell, K. and Sarazynski, A. (2005) "The fundamental challenge in measuring sprawl: which land should be considered", *Professional Geographer* 57: 94–105; Lang, R. E. (2003) "Open bounded places: does the American West's arid landscape yield dense metropolitan growth?" *Housing Policy Debate* 13: 755–778.

105. Frumkin, H., Frank, L. and Jackson, R. (2004) *Urban Sprawl and Public Health,* Washington DC: Island Press.

106. Kahn, M. E. (2000) "The environmental impact of suburbanization", *Journal of Policy Analysis and Management,* 19: 569–586.

107. Volstad, J. H., Roth, N. E., Mercurio, G., Southerland, M. T. and Strebel, D. E. (2003) "Using environmental stressor information to predict the ecological status of Maryland non-tidal streams as measured by biological indicators", *Environmental Monitoring and Assessment* 84: 219–242.

108. Hardoy, J., Mitlin, D. and Satterthwaite, D. (2001) *Enviromental Problems in an Urbanizing World.* London and sterling VA: Earthscan. p. 101.

109. There are numerous names for shanty settlements or squatter settlements. They are called "ranchos" in Venezuela, "pueblo joven" in Peru, "favelas" in Brazil, "barong–barongs" in the Philippines and "kevettits" in Burma. The United Nations uses the term "slum".

110. Brunn, S., Williams, J. and Ziegler, D. (eds) (2003) *Cities of the World: World Regional Urban Development,* New York: Rowman and Littlefield. p. 18.

111. United Nations Centre for Human Settlements, (2001) *Cities in a Globalizing World: Global Report on Human Settlements, 2001.* London and Sterling, VA: Earthscan, p xxvi.

112. Hardoy, J., Mitlin, D and Satterthwaite, D. (2001) p. 42–44.

113. Hardoy, J., Mitlin, D and Satterthwaite, D. (2001) p. 42.

114. Hardoy, J., Mitlin, D and Satterthwaite, D. (2001) p. 6.

115. Hardoy, J., Mitlin, D and Satterthwaite, D. (2001) p. 7.

116. O'Hare, G. and Barke, M. (2002) "The favelas of Rio de Janeiro: a temporal and spatial analysis", *GeoJournal* 56(3): 225–240.

117. The World Health Organization, (1992) *Our Planet, Our Health,* Report of the Commission on Health and Environment, Geneva, p. 204.

118. Hardoy, J., Mitlin, D and Satterthwaite, D. (2001) p. 72.

119. Hardoy, J., Mitlin, D and Satterthwaite, D. (2001) p. 1–2.

Chapter 5 Urban sites

120. Colten, C. (2004) *An Unnatural Metropolis; Wresting New Orleans from Nature,* Baton Rouge: LSU Press; Lewis, P. F. (2003, 2nd edn) *New Orleans: The Making of an Urban Landscape,* Sante Fe: Center for American Places.

121. Banham, R. (1971) *Los Angeles,* Harmondsworth: Penguin.

122. Harvey, D. (2003) *Paris: Capital of Modernity,* London: Routledge.

123. Appadurai, A. (2000) "Spectral housing and urban cleansing: notes on millennial Mumbai", *Public Culture* 12: 627–651. See also Mehta, S. (2005) *Maximum City: Bombay Lost and Found,* New York: Vintage.

124. Meyer, W. B. (1994) "Bringing hypsography back in: altitude and residence in American cities", *Urban Geography* 15: 505–513.

125. Patz, J. A., Martens W. J., Focks D. A. and Jetten T. H. (1998) "Dengue fever epidemic potential as projected by general circulation models of global climate change", *Environmental Health Perspective,* 106:147–153.

126. Osborne, M. E. (2000) *The Mekong: Turbulent Past, Uncertain Future,* New York: Atlantic Monthly Press.

127. Gumprecht, B. (1999) *The Los Angeles River; the Life, Death and Possible Rebirth,* Baltimore and London: Johns Hopkins University Press.

Chapter 6 Cities, environmental hazards and disasters

128. (http://www.gpoacess.gov/congress/index/html (2006) A failure of Initiative: Final Report of the Select Bipartisan Committee to Investigate the Preparation for and response to Hurricane Katrina, US Government Printing Office, Washington, DC.

129. Hartman, C. and Squires, G. (eds) (2006) *There is No Such Thing as a Natural Disaster,* New York: Routledge.

130. Eakin, H. and Luers, A. L. (2006) Assessing the vulnerability of social-environmental systems, *Annual Review of Environment and Resources* 31: 365–394. See also Cutter, S. L., Baruff, B. T. and Shirley, W. L. (2003) "Social vulnerability to environmental hazards", *Social Science Quarterly* 84: 242–261.

131. Chafe, Z. (2007) "Reducing natural disaster risk in cities", *In State of The World: Our Urban Future,* New York: Norton, pp. 112-129; see also, Bull–Kamanga, L., Diagne, K., Lavell, A., Leon, E., Lerise, F., MacGregor, H., Maskrey, A., Meshack, M., Pelling, M., Reid, H., Satterthwaite, D., Songsore, J., Westgate, K. and Yitambe, A. (2003) "From everyday hazards to disasters: the accumulation of risk in urban area", *Environment and Urbanization* 15:193–203.

132. Davis, M. (1998) *The Ecology of Fear,* New York: Holt.

133. Pelling, M. (2003) *The Vulnerability of Cities: Natural Disaster and Social Resilience,* New York and London: Routledge.

134. Tinniswood, A. (2003) *By Permission of Heaven: The True Story of The Great Fire of London,* London: Jonathan Cape.

135. Sawislak, K. (1995) *Smoldering City; Chicagoans and the Great Fire, 1871–1874,* Chicago: University of Chicago Press.

136. (http://www.nationalcenter.org/ChicagoFire.html)

137. Maps of the fires can be found at http://map.sdsu.edu

138. Barry, J. (1997) *Rising Tide: The Great Mississippi Flood of 1927 and How It Changed America,* New York: Simon and Schuster.

139. Kim, K.G. (1999) "Flood hazard in Seoul: a preliminary assessment", Mitchell, J. K. (ed.) *Crucibles of Hazards: Mega-cities and Disasters in Transition,* Shibuya–ku: United Nations University Press.

140. (http://www.1906eqconf.org/mediadocs/BigonestrikesReport.pdf). Accessed July 2006.

141. Davis, D. (2005) "Reverberations: Mexico City's 1965. earthquake and the transformation of the capital", in Vale, L. J. and Campanella, T. J. (eds) *The Resilient City: How Modern Cities Recover From Disasters,* New York: Oxford University Press.

142. Davis, D. (2005), p. 276.

143. Chen, B. (2005) "Resist the earthquake and rescue ourselves: the reconstruction of Tangshan after the 1976 earthquake", in Vale, L. J. and Campanella, T. J. (eds) *The Resilient City: How Modern Cities Recover From Disasters,* New York: Oxford University Press, p. 251.

144. Vale, L. J. and Campanella, T. J. (eds) (2005) *The Resilient City: How Modern Cities Recover From Disasters,* New York: Oxford University Press.

145. Fallahi. A. (2006) "Mobilization of local and regional capabilities through community participation in the process of recovery after the Bam earthquake in Iran". Paper presented to the International Geographical Union Conference, Brisbane, Australia, 5 July.

Chapter 7 Urban ecology

146. Pickett, S. T. A., Burch, W. R. Jr., Dalton, S. D., and Foresman, T. W. (1997) "Integrated urban ecosystem research", *Urban Ecosystems* 1: 183–184.
147. Cadenasso, M., Pickett, S. T. A. and Grove, M. J. (2006) "Integrative approaches to investigating human-natural systems: the Baltimore ecosystem study", *Natures Sciences Sociétés* 14: 4–14.
148. Georgi, N. J. and Zafiriadis, K. (2006) "The impact of park trees on microclimates in urban areas," *Urban Ecosystems* 10:195–209.
149. Wrigley, N. (2002) "Food deserts in British cities: policy context and research practice", *Urban Studies* 39: 2029–2040.
150. For a detailed discussion of the specific measures see the website http://www.footprintnetwork.org/
151. Wackernagel, M. and Rees, W. (1996) *Our Ecological Footprint: Reducing Human Impact on The Earth,* Gabriola Island, BC: New Society Publishers.
152. Geis, S. (2006) "Northern Nevadans don't want to gamble with their water", *The Washington Post,* August 15, p. A3.
153. Pearce, D. and Barboer, E. (2000) *Blueprint for a Sustainable Economy,* London: Earthscan.
154. Jenkins, J., Riemann, R., Groffman, P., Pouyat, R. V., Grove, J. M., Nowak, D. and Birdsey, R. (2001) "What do urbanized regions contribute to the global C balance?" American Geophysical Union, Spring Meeting. Abstract #B42A–11; Pouyat, R. V., Yesilonis, I. D. and Nowak, D. J. (2006) "Carbon storage by urban soils in the United States", *Journal of Environmental Quality* 35: 1566–1575.
155. Robbins, P. and Sharp, J. (2003) "The lawn-chemical economy and its discontents", *Antipode* 35: 955–979.
156. Groffman, P. M., Law, N. L., Belt, K. T., Band, L. E. and Fisher, G. T. (2004) "Nitrogen fluxes and retention in urban watershed ecosystems", *Ecosystems* 7: 393–403.
157. Blewett, C. M. and Marzluff, J. M. (2005) "Effects of urban sprawl on snags and the abundance and productivity of cavity-nesting birds", *The Condor* 107: 678–693.
158. Blair, R. (2004) "The effects of urban sprawl on birds at multiple levels of biological organization", *Ecology and Society* 9: 1–21.
159. Gregg, J. W., Jones, C. G. and Dawson, T. E. (2003) "Urbanization effects on tree growth in the vicinity of New York City", *Nature* 424: 183–187.
160. Hope, D., Gries, C., Zhu, W., Fagan, W., Redman, C., Grimm, N. B., Nelson, A. L., Martin, C. and Kinzig, A. (2003) "Socioeconomics drive urban plant diversity", *Proceedings of the National Academy of Sciences of the USA* 100: 8788–8792.
161. Chai, Y., Zhu, W. and Han, H. (2002) "Dust removal effect of urban tree species in Harbin", *Ying Yong Sheng Tai Xue Bao* 13:1121–1126.
162. For a fuller discussion of the context of the Chicago School see Short, J. R. (2006) *Urban Theory,* Houndmills and New York: Palgrave.

163. http://www.umbc.edu/ges/research/sohn%27s_res_info/sohn_figure3.htm Accessed July 2006.

164. A recent study of gardens, in Hobart, Tasmania is available in Kirkpatrick, J.B. (2006) *The Ecologies of Paradise: Explaining the Garden Next Door*, Hobart: Pandani Press.

165. Cielsewicz, D. (2002) "The environmental impacts of sprawl," in G. Squires (ed.) *Urban Sprawl: Causes, Consequences and Policy Responses,* Washington, DC: Urban Institute Press.

166. Volstad, J. H., Roth, N. E., Mercurio, G., Southerland, M. T. and Strebel, D. E. (2003) "Using environmental stressor information to predict the ecological status of Maryland non-tidal streams as measured by biological indicators", *Environmental Monitoring and Assessment* 84: 219–242.

Chapter 8 Water pollution and the city

167. Lee, J.H. and Bang, K. W. (2000) "Characterization of urban stormwater runoff", *Water Research* 34(6): 1773–1780. See also Gromarie-Mertz, M.C., Garnaud, S., Gonzalez, A. and Chebbo, G. (1999) "Characteristics of urban runoff pollution in Paris," *Water Science Technology* 39(2): 1–8.

168. Environmental Protection Agency (2006) *Report to Congress on the Impacts and Control of CSOs and SSOs,* Executive Summary p. ES–5.

169. "Clean Water Act Reauthorization: management of non-point source pollution", www.cnie.org/nle/h20–2thml, April 12, 2000.

170. Source: www.epa.gov/owow/cwa/history.htm September, 11, 2002.

171. Uitto, J. and Biswas, A. (eds) (2000) *Water for Urban Areas: Challenges and Perspectives,* Tokyo and New York: United Nations University Press, p. xv.

172. Uitto, J. and Biswas, A. (eds) (2000) p. xiii.

173. Uitto, J. and Biswas, A. (eds) (2000) p. xiii.

174. Pezzoli, K. (2001) *Human Settlements and Planning for Ecological Sustainability,* Boston: The MIT Press, p. 59.

175. Ezcurra, E., Mazari–Hiriart, M., Pisanty, I. and Aguilar, A.G. (1999). *The Basin of Mexico: Critical Environmental Issues and Sustainability,* New York: United Nations University Press.

176. Tortajada–Quiroz, C. (2000) "Water supply and distribution in the metropolitan area of Mexico City", in Uitto, J. and Biswas, A. (eds) (2000) *Water for Urban Areas: Challenges and Perspectives,* Tokyo and New York: United Nations University Press, p. 120.

177. National Research Council (1995) *Mexico City's Water Supply,* Washington, DC: National Academy Press, p. 14.

178. Ezcurra, E., Mazari–Hiriart, M., Pisanty, I. and Aguilar, A.G. (1999).

179. Tortajada–Quiroz, C. (2000) p. 113.

180. Ezcurra, E., Mazari–Hiriart, M., Pisanty, I. and Aguilar, A.G. (1999).

181. Barkin, D. (2004). "Mexico City's water crisis", *NACLA Report on the Americas,* July/August pp. 27–28.

182. Barkin, D. (2004).

183. Stille, A. (1998) "The Ganges' next life", *The New Yorker,* January 19, pp. 58–67. B.O.D. stands for Biochemical Oxygen Demand and is a measure of how much oxygen has been used by micro-organisms in the biodegradation process. A river or body of water with high levels of BOD may be unhealthy for aquatic life and if very high, may asphyxiate fish and other marine organisms.

Chapter 9 Air pollution and the city

184. US Environmental Protection Agency (2002) "Latest findings on National Air Quality: 2000 status and trends" at http://www.epa.gov/airlinks

185. Ozone in the stratosphere is naturally occurring and beneficial in that it forms a protective layer that filters out the sun's harmful ultraviolet (UV) radiation. In contrast, ground–level ozone is not beneficial and negatively impacts on human health and the environment.

186. US Environmental Protection Agency (2002).

187. World Bank (2002) *Cities on the move: A World Bank Urban Transport Strategy Review,* Washington, DC: The World Bank.

188. Ostro, B. (2004) *Outdoor Air Pollution: Assessing the Environmental Burden of Diseases at the National and Local Level,* Geneva: World Health Organization (World Health Organization Environmental Burden of Disease Series, No. 5).

189. Elsom, D. (1996) *Smog Alert: Managing Urban Air Quality,* London and Sterling, VA: Earthscan. p. 27.

190. World Resources Institute (1998) *World Resources 1998–99,* World Resources Institute: Washington, DC.

191. Elsom, D. (1996) p. 5.

192. In 1976 the US began phasing out the use of lead as a gasoline additive. By 1986, leaded gasoline was banned in the US. But other countries, particularly in the developing world have delayed these bans, as unleaded gasoline is more expensive.

By 2000 only 42 countries had completely phased out leaded gasoline. A dozen others, including India, were committed to making the switch by 2005. Yet, more than 150 countries have still not decided to mandate the change.

193. The World Bank (1998) "Clean air initiative in sub-Saharan Africa cities: Home" at http://www.worldbank.org/wbi/cleanair/caiafrica/index.htm on February 2003.

194. The industrialized north faces indoor air pollution as well, primarily from chemicals and compounds such as radon or carbon monoxide and tobacco smoke.

195. World Health Organization (2006) "Indoor air pollution fact sheet" retrieved at http://www.who.org.int/mediacenter/factsheets/en/ in July, 2007.

196. The World Resources Institute (1999) "Rising energy use: health effects of air pollution", www.wri.org Accessed Februry 27 2003.

197. World Resources Institute, (1998) *World Resources 1998–99,* World Resources Institute: Washington, DC.

Chapter 10 Garbage and the city

198. Melosi, M. (2000) *The Sanitary City: urban infrastructure in America from colonial times to the present,* Baltimore: The Johns Hopkins University Press, p. 339.

199. Melosi, M. (2000) p. 340.

200. Melosi, M. (2000) p. 338.

201. Urban Development Unit, World Bank (1999) "What a waste: solid waste management in Asia", Washington, DC: World Bank, May, p. 18.

202. Melosi, M. (2000) p. 343.

203. http://www.epa.gov/msw/facts.htm retrieved on September 12, 2006.

204. US EPA (1999) *National Source Reduction Characterization Report for Municipal Solid Waste in the United States,* EPA 530–R–99–034. Retrieved at www.epa.gov/osw on February, 2000, p. 15.

205. Lipton, E. (2000) "Efforts to close Fresh Kills are taking unforeseen tolls", in The *New York Times,* February 21, 2000, Section A, p. 1.

206. Melosi, M. (2000) p. 261.

207. Royte, E. (2005) *Garbage Land: On the Secret Trail of Trash,* New York: Little Brown.

208. Fishbein, B. and Azilmi, S. (1994) *Germany, Garbage and the Green Dot: Challenging the Throwaway Society,* New York: Inform, pp. 18–21.

209. US EPA (1999) *National Source Reduction Characterization Report for Municipal Solid Waste in the United States,* EPA 530–R–99–034. Retrieved at www.epa.gov/osw on February, 2000, p. 14.

210. Urban Development Unit, World Bank (1999) pp. 7–8.

211. Urban Development Unit, World Bank, (1999) p. 3.

212. Urban Development Unit, World Bank, (1999) pp. 10–11.

213. U.S. Department of Commerce, (2001) *Brazil: Solid Waste Statistical Data,* STAT–USA on the Internet. Retrieved at http://srategis.ic.gc.ca/SSG/ dd72515e.html on September 25, 2006.

Chapter 11 Race, class and environmental justice

214. Cutter, S. (1995) "Race, class and environmental justice", *Progress in Human Geography* 19: 107–118.

215. United Church of Christ. Commission for Racial Justice (1987) *Toxic Wastes and Race in the United States: A National Report on The Racial and Socio-economic Characteristics of Communities With Hazardous Waste Sites,* New York: Public Data Access: Inquiries to The Commission.

216. Boone, C. G. and Modarres, A. (1999) "Creating a toxic neighborhood in Los Angles County", *Urban Affairs Review* 35: 163–187. See also Pulido, L. (2000) "Rethinking environmental racism: white privilege and urban development in Southern California", *Annals of the Association of American Geographers* 90: 12–40.

217. Bullard, R. (1993) *Confronting Environmental Racism: Voices from The Grassroots,* Boston: South End Press. See also Bullard, R. D. and Johnson, G. S. (2000) "Environmental justice: grassroots activism and its impact on public policy deci-sion making", *Journal of Social Issues* 56: 555–578.

218. Brulle, R, J. and Pellow, D. W. (2006) "Environmental justice: human health and environmental inequalities", *Annual Review of Public Health* 27:103–124.

219. Crenson, M. (1971) *The Un-politics of Air Pollution: A Study of Non-Decisionmaking in Cities,* Baltimore: Johns Hopkins University Press; see also Benton, L. M. and Short, J. R. (1999) *Environmental Discourse and Practice,* Oxford: Blackwell, Chapter 6.

220. Gibbs, L. (1993) Foreword, in R, Hofrichter (ed.) *Toxic Struggles; the Theory and Practice of Environmental Justice,* Philadelphia: New Society Publishers.

221. See Short, J. R. (1989) *The Humane City,* Oxford: Blackwell, especially chapter 4, "Cities as if only some people matter".

222. Kenzer, M. (2000) "Healthy cities; a guide to the literature", *Public Health Reports* 115: 279–289; Hartley, N. and Wood, C. (2005) "Public participation in environ-mental impact assessment; implementing the Aarhus Convention", *Environmental Impact Assessment Review* 25: 319–340.

Chapter 12. Sustainable urban development

223. As quoted by Beatley, T. (2007) "Sustaining the city: urban ecology", paper presented at *Symposium on Framing a Capital City*, National Building Museum, Washington, DC, April 11, 2007.

224. Singer, M. (2003) "The haves and the haves", *The New Yorker*, August 11: 56–61.

225. From the charter of the New Urbanism:

> Neighborhoods should be compact, pedestrian-friendly, and mixed use. Many activities of daily living should occur within walking distance, allowing independence to those who do not drive, especially the elderly and the young. Interconnected networks of streets should be designed to encourage walking, reduce the number of and length of automobile trips and conserve energy. Within neighborhoods a broad range of housing types and price levels can bring people of diverse races and incomes into daily interaction, strengthening the personal and civic bonds essential to an authentic community.
>
> <http://www.cnu.org/charter.html> (9 November, 1999)

226. Frantz, D. and Collins, C. (1999) *Celebration USA*, New York: Henry Holt. See also Ross, A. (1999) *The Celebration Chronicles*, New York: Ballantine.

227. Girardet, H. (2004) *Cities People Planet: Livable Cities for a Sustainable World*, Chichester: Wiley Academy.

228. For a range of opinion see Talen, E. (1999) "Sense of community and neighborhood form: an assessment of the social doctrine of New Urbanism", *Urban Studies* 36(8): 1361–1379; Krieger, A. (1998) "Whose Urbanism?" *Architecture*, (November): 73–77; and Ford. L. (1999) "Lynch revisited: New Urbanism and theories of good city form", *Cities* 16: 257–277.

229. Mayer, H. and Knox, P. L. (2006) "Slow cities; sustainable places in a fast world", *Journal of Urban Affairs* 28: 321–334.

230. Tyler, N. (2000) *Historic Preservation: An Introduction to Its History, Principles and Practice*, New York: W.W. Norton.

231. Tyler, N. (2000) pp. 23–24.

232. Tyler, N. (2000) pp. 27–28.

233. For example, the United Nations Commission on Environment and Development released the report "Our Common Future" in 1987, and in 1992 the UN Rio de Janeiro "Earth Summit" conference called for "sustainable development". The term and concepts thus entered the international mainstream.

234. Wheeler, S. (2004) *Planning for Sustainability: Creating Livable, Equitable and Ecological Communities*, London and New York: Routledge, p. 23.

235. United Nations, World Commission on Environment and Development (1987) *Our Common Future,* p. 8.

236. Wheeler. S. (2004), pp. 23–25.

237. Spirn, A.W. (1985) "Urban nature and human design: renewing the Great Tradition", *Journal of Planning Education and Research* 5: 475–495.

238. Low, N., Gleeson, B., Green, R. and Radović D. (2005) *The Green City: Sustainable Homes, Sustainable Suburbs.* Oxford: Routledge, p. 44.

239. http://hortweb.cas.psu.edu/research/greenroofcenter/history.html

240. Low, N., Gleeson, B. Green, R. and Radović, D. (2005) p. 54.

241. Low, N., Gleeson, B., Green, R. and Radović, D. (2005) pp. 78–79.

242. Low, N., Gleeson, B., Green, R. and Radović, D. (2005) p. 81.

243. Ulrich, R.S. (1984) "Views through a window may influence recovery from surgery", *Science* 224: 420–421.

244. Girardet, H. (2004) p. 239.

245. Girardet, H (2004) p. 247.

246. Low, N., Gleeson, B., Green, R. and Radović, D. (2005) p. 135.

247. Low, N., Gleeson, B., Green, R. and Radović, D. (2005) p. 138.

248. Low, N., Gleeson, B., Green, R. and Radović, D. (2005) p. 149.

249. Girardet, H. (2004) p. 131.

250. Girardet, H. (2004) p. 136.

251. Brajer, V. and Mead, R.W. (2003) "Blue skies in Beijing?: Looking at the Olympic effect", *Journal of Environment and Development* 12(2): 239–263.

252. Rabinovitch, J. (1997) "A success story of urban planning: Curitiba", in U. Kirdar, (ed.) *Cities Fit for People,* New York: United Nations Press, p. 425.

253. Rabinovitch, J. (1997) p. 424.

254. Rabinovitch, J. (1997) p. 424–426.

255. Rabinovitch, J. (1997) p. 429.

256. Irazábal, C. (2005) *"City Making and Urban Governance in the Americas: Curitiba and Portland,* Burlington: Ashgate". See also Gilbert, R. (1996) *Making Cities Work: the Role of Local Authorities in the Urban Environment,* London: Earthscan.

257. While, A., Jonas, A.E.G. and Gibbs, D. (2004) "The environment and the entrepreneurial city: searching for the urban 'sustainability fix' in Manchester and Leeds", *International Journal of Urban and Regional Research* 28(3): 549–569.

Index